Modern
Bacterial
Taxonomy

Modern Bacterial Taxonomy

Brian Austin
and
Fergus Priest
Department of Brewing and Biological Sciences
Heriot-Watt University, Edinburgh

 Van Nostrand Reinhold (UK)

Published in 1986 by
Van Nostrand Reinhold (UK) Co. Ltd
Molly Millars Lane, Wokingham, Berkshire, England

Photoset in Plantin 10 on 11pt by Kelly Typesetting Limited
Bradford-on-Avon, Wiltshire

Printed in Great Britain by
Billing & Sons Ltd, Worcester

ISBN 0 442 31736 0

Contents

Preface

This book is intended to cater for the need for a concise, easy to understand, relatively inexpensive text, suitable for undergraduate students who receive integrated courses in bacterial taxonomy. In addition, there is a perceived need for such a text for postgraduate and research scientists of other disciplines who encounter and use bacterial taxonomy as a peripheral subject but may find the current methods and range of the subject daunting. Finally, we hope that it will be read by all microbiologists who have a passing interest in the subject.

The scope of the book covers all aspects of the theory and practice of bacterial taxonomy. Starting with a historical view on the significant developments, the reader is directed through accounts of the philosophy of classification, i.e., phenetic and phylogenetic the modern approaches of chemotaxonomy and numerical taxonomy, the intricacies of cladistics, nomenclature and identification methods, and thence to the possible interactions with allied disciplines, namely biotechnology, ecology, genetics and pathology. The role of microbial culture collections is also included.

The need for such a book was conceived after many years of teaching undergraduate and postgraduate courses in bacterial taxonomy. Students could be directed towards a range of specialist books and chapters dealing with certain aspects of taxonomy or towards the elementary coverage of some well established basic bacteriological texts. Unfortunately, a single text devoted to the subject of bacterial taxonomy was absent from bookshelves.

In a book such as this, which deals with diverse disciplines ranging from molecular biology of nucleic acids to multivariate statistics, inevitably some parts will be condensed and read more as a review article or monograph than a student textbook. This approach was considered preferable to the alternatives, which were either to produce a much longer, and consequently more expensive book, of which there are several available, or to simply omit some topics that we considered were very important. We therefore apologise if the reader finds some chapters to be too densely written or perhaps awkward questions avoided, but we emphasize that our first priority was to provide a concise, yet comprehensive, text.

We are particularly grateful to the helpful co-operation of our publisher, Dominic Recaldin, and to the excellent secretarial assistance of Mrs Mairie Russell. It is our earnest belief that if the book stimulates interest in bacterial taxonomy among the readers, then the task will have been worthwhile.

Introduction

There has long been a fascination with the relationships among living things and their arrangement into categories, beginning with the writings of Aristotle and culminating in the great naturalists of the last century.

The formulation of relationships was accepted without question, largely because living organisms could be readily distinguished by morphology. It may seem obvious that a dog is distinct from a horse, and although the reasons may be difficult to explain, they will include assessment of size and shape differences. In the case of bacteria, the problem is exaggerated because of the extremely small size of the organisms and lack of pronounced morphological variation. It was indeed astonishing that van Leeuwenhoek, the founding father of microbiology, managed to observe what are now regarded as bacteria-like objects, considering that the maximum magnification of his simple microscope did not exceed 300 times. The 'animalcules', recorded in his landmark publication of 1684, constituted the first legitimate report of microbes (bacteria?) and he made careful illustrations that, to some extent, suggested morphological variations between the cells. Nevertheless, it took many more years for classifications of bacteria to be developed. Perhaps some of the delay stemmed from a genuine lack of interest in taxonomy. Much of the early microbiology was conducted by medical scientists; Pasteur, for example, was more interested in the physiology and pathogenicity of micro-organisms than in their relationships to each other. Unfortunately, this attitude still prevails with some scientists.

Early classifications were based on largely morphological information, as determined by light microscopy. Indeed, the different morphological forms of bacteria were described lucidly by Müller in the eighteenth century. Thereafter, credit must be given to Ehrenberg and Dujardin for their seemingly revolutionary classifications. For example in the 1830s before the advent of pure culture techniques, Ehrenberg described the family '*Vibrionia*' and divided it into four genera.

Following the interest in morphology, pure culture techniques allowed more rigorous study of bacterial metabolism and physiological information was included in classifications. A classic example is the work of Orla Jensen, who in 1919 divided the lactic acid bacteria into four genera which, with modifications, are used today. Thus *Lactobacillus*,

Streptococcus, '*Betacoccus*' (now *Leuconostoc*) and '*Tetracoccus*' (now *Pediococcus*) were classified on the basis of physiological (type of fermentation, growth temperatures, etc.) and morphological criteria. A particularly readable account of these early contributions to microbiology has been published by T.D. Brock (*Milestones in Microbiology*, American Society for Microbiology, Washington, D.C., 1975).

More recently the trend has been to supplement the physiological information with data from chemical analyses and molecular biology. This book provides an introduction to bacterial classifications with particular emphasis on modern trends in the subject and the use of numerical techniques to quantify relationships based on morphology, physiology, chemical structure and the molecular biology of bacteria. It also covers recent approaches to the identification of bacteria. However, before considering the background to bacterial systematics in more detail, it will avoid confusion to define some key terms.

Taxonomy (often considered to be synonymous with classification) is regarded as the theory of '*classification*' (this concerns the arranging of organisms into groups); '*taxon*' (pl. taxa) is a group of individuals of any rank. Genera, families and species are taxa. *Nomenclature* is the process of allocating names to the taxa. *Identification* is the means by which unknown organisms are allocated to previously described taxa. All these terms apply to *systematics*, which is the study of the diversity and relationships among organisms.

1 Classification

What are the purposes of classifications? There are, in fact, at least three important answers to this question. The first concerns the transfer of information. Quite simply, a classification is a means of summarizing and cataloguing information about organisms. It follows that a classification is predictive. For example, the names *Escherichia coli* or *Bacillus subtilis* refer to groups of individuals, about which a great deal is known.

A second purpose for classification is that organisms must be categorized such that the identification systems may be devised for the recognition of new isolates. It is evident that without prior arrangements of individuals into groups it would be impossible to assign new isolates to a taxon.

The third purpose of classification may be considered to provide an insight into the evolutionary pathways of bacteria.

If classifications are to serve these purposes effectively, they should fulfil several criteria. Firstly, a classification should have high information content: essentially the greater the amount of information on which it is based, then the greater will be the 'predictivity' of the classification and the more generalizations that may be made about the taxa involved. Secondly, classifications should be stable. This may seem obvious, but a classification, in which the composition and descriptions of taxa change frequently, is confusing and unhelpful. Finally, it is important that the development of a classification should be empirical, reproducible and scientifically based.

1.1 DEFICIENCIES OF TRADITIONAL CLASSIFICATIONS

It shall be apparent to students of microbiology that bacterial classifications have, until recently, failed to achieve the necessary requirements of predictivity, stability and objectivity. One reason for this is that these aspects of a classification are intimately entwined and failure to achieve one is generally reflected in failure to satisfy all three. It is pertinent to enquire precisely what the early bacterial taxonomists were doing wrong. These shortcomings have been fully discussed by Cain (1963), but need to be briefly reviewed here because it is important to understand the reasons for the failure of early classifications.

Classifications have been traditionally based on Linnaeus's principles, which suggested that the process of classification should be conducted from 'above' and, by starting from the overall groups encompassing all living things, repeated divisions could be made until the species level was reached. In this system, species were recognized as being indivisible, i.e., the basic taxonomic unit, and at every rank, taxa were defined by specific features, which reflected the 'essential nature' of the group. The taxonomist had simply to discover these features to effect classification. This is regarded as an *a priori* choice of characters, since it supposes that the important feature(s) of a group can be chosen deductively. However, this is purely subjective, since it cannot be known by intuition which features best reflect the 'essential nature' of the group or, for that matter, the organism. When such 'important' features are discovered, the reason is usually that the group has been already subjected to systematic study and useful diagnostic characters have been highlighted. The *a priori* choice of characters initially leads to serious problems because of disagreement between scientists. Characters considered by one worker to be of inestimable importance for defining groups may be totally disregarded by others. Thus, a lengthy discourse occurred earlier this century concerning the relative importance of morphological characters, such as flagella pattern, cell shape, and physiological characters for classification. Indeed, this approach was taken to the extreme in the misguided assumption that a progression of characters existed that defined the heirarchy of taxa. It was considered that morphological characters defined genera, physiological characters defined species and serological features could be used to define sub-species.

If the criteria of a good classification as discussed above are considered, it will be apparent that the traditional approach failed in all respects. Since only a few so-called 'important' characters were used to construct the classification, it was based on little information and lacked predictivity. Very few assertions could be made about the taxa. Secondly, the classifications were unstable because the choice of the important characters was subjective. Different taxonomists expressed contrasting views about the composition and defining features of taxa. This resulted in the continual revision of taxa with new descriptions and quite often new names. Finally, it may be argued that classification has not been conducted as an empirical science, because of reliance on subjectivity and intuition in the choice of defining characters. It was not repeatable because of the involvement of personal judgements by the scientists.

Having argued against the traditional approach to classification, it is necessary to provide a satisfactory alternative, but to precede this, consideration must be given to the kinds of classification that are available.

1.2 THE RANGE OF CLASSIFICATIONS

It is important to emphasize at this point that there is no single unifying classification of living organisms. Biologists are dogmatic in their belief that there is a single correct way to classify the individuals of a population. Sneath (1983) suggested that 'biology is so complex that underlying explanations are very difficult to detect: therefore there is a demand for a single dominating concept that will encompass all of its phenomena'. Alternatively, a view has been expressed that life may have started as a single event (whether caused by Creation or natural process) and consequently there has been a tendency to think of a single plan underlying all living organisms that can be used as a basis for classification. However, there is no reason to require a single classification, since classifications are devised by man for various purposes. Since we have many purposes in mind there are many types of classification, and it is possible to classify them! Essentially there are three types of classification dependent on the nature of the relationships used in their construction.

1.2.1. Special-purpose classifications

Bacteriologists often use classification and identification schemes designed for their particular discipline. For example, food microbiologists or insect pathologists might use specific identification systems for the bacteria that they are most likely to encounter. Justifiably, these schemes ignore all other bacteria as irrelevant. They may be extremely useful for the specialist microbiologist, but are of little value to microbiology in general because most bacteria are excluded.

Special-purpose classifications are artificial in that they seldom display the 'natural' relationships among the organisms. The distinction between *Shigella dysenteriae* and *E. coli* is a well known example. Strains of these taxa share considerable DNA sequence homology and are phenotypically very similar. From virtually every viewpoint they could be considered as a single species. However, the more serious pathogenicity of *S. dysenteriae* is of considerable importance to the clinician, and consequently, the separate taxa have been retained. Similarly, *Bacillus cereus* and *B. thuringiensis* are virtually identical, but separate species status has been maintained for *B. thuringiensis*, which contains the crystalliferous insect pathogenic bacilli. These artificial divisions obviously have their uses, and in the two examples cited above the needs of the specialist have been recognized by microbiologists in general and these opinions are incorporated into the mainstream of microbiological classification. It is necessary for the clinical bacteriologist to distinguish *S. dysenteriae* from *E. coli*, and for the insect pathologist to recognize the insect pathogenic bacilli; but we must also recognize the limitations of this approach.

Artificial classifications are monothetic, in that a single feature (pathogenicity in the above examples) is deemed both sufficient and necessary for the placement of an organism in a group. Monothetic classifications suffer from the serious disadvantage noted for traditional classifications in Section 1.1. They are based on restricted information and tend to be unstable, since bacteriologists with different interests adopt different schemes. For example, the plant saprophyte *Erwinia herbicola* is synonymous with an intestinal organism, *Enterobacter agglomerans*. Moreover, identification schemes derived from monothetic classifications readily lead to misidentification, since the unknown organism need only be aberrant in one feature to be assigned to the wrong taxon. Relatively non-pathogenic isolates of *S. dysenteriae* would be placed in the genus *Escherichia*. Artificial classifications have their uses, but as a general system for use by all microbiologists their limitations are severe.

1.2.2. Natural (phenetic) classifications

The alternative to the special-purpose classification is the general-purpose classification, a system that is of value to all microbiologists whatever their discipline. Such a classification should encompass all bacteria and all aspects of these bacteria. Since special-purpose classifications are artificial so general-purpose classifications can be described as 'natural'. Natural in this sense can be attributed to Gilmour (1951) and was developed by Sneath (1962) to refer to relationships between organisms based on their overall similarity or affinity. Natural relationships embody all aspects of the organisms from molecular structure through physiology, to habitat. Such relationships are termed *phenetic* and refer to affinities based on the complete organism (genotype and phenotype) as it exists at present with no reference to the evolutionary pathways or ancestry of the organism. This contrasts with the term natural used in its evolutionary context (Section 1.2.3).

In phenetic classifications, organisms are arranged into groups (phena) on the basis of high overall similarity using both phenotypic and genotypic characters. This approach encompasses all measurable features of the organisms so that the resultant classification should be useful to all microbiologists. Moreover, the taxa are polythetic rather than monothetic because they are defined as having a high number of features in common and there is no requirement for the presence of a particular attribute. Thus, individuals aberrant in a specific character can be accommodated by such groups. In a phenetic classification, *E. coli* and *S. dysenteriae* would be placed in the same species, since they have high genotypic and phenotypic similarity. Consequently, the distinction attributed to relative pathogenicity would not be given undue importance. This describes the natural relationships of these bacteria, and, although it may alarm the clinical bacteriologist, it is a sensible approach insofar as strains of *E. coli* carrying toxin and surface antigen-encoding plasmids are associated with serious diarrhoeal disease. This fact is often obscured by the artificial division of these organisms into 'pathogenic' and 'non-pathogenic' types.

It follows that phenetic classifications do not suffer the shortcomings of their artificial counterparts. Because the classification is based on the overall properties of the organisms it has a high information content with the associated predictivity. The classification is also more stable since the same names should be used to describe the same taxa regardless of the interests of the microbiologist. Finally, since phenetic classifications are always, in theory, polythetic, identification should be more accurate since individuals are assigned to groups using several characteristics and organisms that do not fully conform are still accommodated by such groups.

1.2.3. Natural (phylogenetic) classifications

For many biologists, particuarly those who study animals and plants, the term 'natural' refers to an intrinsic quality of nature and natural groups are 'species or groups of species that exist in nature as a result of a unique history of descent with modification' (i.e. evolution) (Wiley, 1981) or 'the true historical entities produced by the evolutionary process' (Cracraft, 1983). These natural classifications are based on *phylogenetic* (genealogical) relationships in that they attempt to trace the evolutionary pathways that have given rise to the organisms as we view them today, with the classification exactly reflecting the line of ancestry. This classification will be congruent with the phenetic classification if there has been no parallel or convergent evolution, but the two classifications will differ if convergent evolution or recent gene transfer gives rise to organisms that are phenetically similar but have different ancestry (Fig. 1.1). Much of what follows is more relevant to plant or animal biologists than micro-

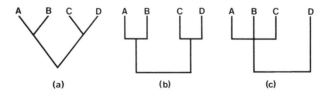

Fig. 1.1 Three dendrograms representing (a) a cladogram indicating the phylogenetic relationships of taxa A,B,C,D, (b) a phenogram of the same taxa in which evolution has been assumed to be divergent and at constant rate and (c) a phenogram in which convergent evolution or recent gene transfer has resulted in C being phenetically related to A,B.

biologists, but with the advent of macromolecule sequencing, bacteria
are becoming more amenable to phylogenetic analysis and therefore the
major schools of phylogenetic thinking are described briefly below.

Amongst phylogeneticists, there is considerable controversy. The
term *cladistic* refers to the branching pattern that describes the pathway
of ancestry of a group of organisms. Hennig (1966) showed how cladistic
relationships might be inferred. Hennigean cladistics involves the determination of monophyletic groups that can be defined as groups distinguished by a set of characters inherited from an ancestor. In other
words, all members of a monophyletic group possess a homologous
character either in its primitive (earlier) or derived (later) form and it is
the joint possession of this character by all descendants of the species that
defines the group. For example, in Fig. 1.2 BC is a monophyletic group

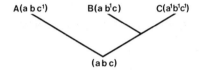

Fig. 1.2 A cladogram of species A,B and C. B and C constitute a monophyletic
group defined by the shared derived character b[1]. A and B are not monophyletic
because species C is excluded. Species A and C are not monophyletic because
they have acquired character c[1] independently; it is not a homologous character.
(After Williams, 1985.)

defined by the shared character b[1]. A and B are not monophyletic
because C is excluded from the group based on the shared primitive
character a. Moreover, A and C are not monophyletic because the only
character which they have in common was acquired independently, i.e.,
is non-homologous. It is argued that by careful selection of characters,
non-convergent homologous characters and monophyletic groups can be
determined and built into a heirarchy that must reflect the evolutionary
pathway.

'Traditional evolutionists', however, follow the ideas of Simpson and
Mayr and dilute phylogenetic relationships with an element of phenetic
similarity when constructing the classification. Classification is practised
with reference to the phylogeny, but without the requirement that all
groups be monophyletic. The difference between Simpsonian and
Hennigean cladistics has been neatly summarized by Williams (1985);
'According to Simpson (1961) the classification should be consistent with
the presumed phylogenetic relationships, whereas for Hennig they are
one and the same thing'.

To further confuse the issue, 'transformed cladists' or natural order
systematists have drifted from Hennig's approach to such an extent that
they are more akin to pheneticists and their views have no connection

with phylogeny. They argue that evolution cannot be known and evolutionary homologous characters are unknown and can only be deduced from the classification itself. As emphasized by Sneath (1983) 'Homologies cannot be recognized from the way character states are distributed in monophyletic groups *because these groups cannot be constructed until the homologies have first been recognized*'. The transformed cladists approach is therefore to assess the distribution of various character states among a group of organisms and to arrange the organisms into a classification determined by just one criterion, maximum parsimony, that is the route which involves the minimal number of changes to arrive at the simplest possible arrangement (see Chapter 4.) Thus the 'natural order' of taxa is revealed without recourse to any evolutionary theories.

1.3 MERITS OF PHENETIC VERSUS PHYLOGENETIC CLASSIFICATIONS

Since it is commonly (but incorrectly) upheld by taxonomists that there is only a single 'natural' classification of organisms, there is obviously considerable controversy among the different schools of thought as to which approach should be adopted. It will be useful, therefore, to consider the relative merits of phenetics and phylogenetics.

1.3.1 'Goodness' of the classification

The basic criterion of the phylogenetic classification is that it should precisely reflect the evolutionary pathway of the organisms. However, it is impossible to compare the cladogram with the true cladogeny because the latter is unknown (merely inferred from the cladogram).

The goal of the pheneticist is less well defined, but it could be argued that the classification should represent, as accurately as possible, the affinities between each and every organism. Various statistical measures have been developed to test the distortion within a heirarchial classification and optimum procedures for constructing a phenetic classification have been proposed (see Chapter 2). However, because of the difficulties in defining the ultimate phenetic classification we must discuss the relative merits of phylogenetic and phenetic classifications on some other grounds.

1.3.2 Verifiability

Since the construction of a classification is a scientific exercise it should be a testable hypothesis or contain testable hypotheses. The phylogenetic approach is not verifiable in this sense because the only way to test a cladogram is with a second cladogram based on a different data set. As pointed out by Sneath (1983), if both cladograms have been derived using the same assumptions they may well be congruent, but this does not test those assumptions, and if different assumptions are used the studies are not comparable. Moreover, all current theories of evolution may seem incorrect to the next generation of biologists.

Conversely, the phenetic classification is verifiable. The operation of producing a phenetic classification involves gathering data that are then analysed using established statistical techniques (see Chapter 2). This process is entirely objective and can be repeated (verified) by a second scientist. Moreover, phenetic classifications are independently testable. As new data are generated by advances in science so they can be included in the classification. If the original classification was correct, in that it accurately represented the affinities of the organisms, the new data should not alter it. If however, the new information does affect the classification it can be included and a composite, improved classification is constructed. Phenetic classifications are therefore verifiable internally and with respect to new data.

1.3.3. Practicalities

One of the principal aims of a classification is to provide a scheme whereby unknown orgnisms may be identified. For the microbiologist, simple, reliable, rapid tests are needed. Phenetic classifications can be analysed to select the most diagnostic characters for delineation of groups and to provide reliable identification schemes (Chapter 6). Phylogenetic classifications of bacteria rely largely on protein or RNA sequence data and are not particularly useful for identification purposes, although new developments in this area will probably revolutionize this aspect of taxonomy (Chapter 6).

1.4 THE CHOICE BETWEEN PHENETIC AND PHYLOGENETIC CLASSIFICATIONS

Despite various claims in the literature, there is no firm evidence that cladistic classifications are any more predictive, congruent or stable than phenetic classifications (Sokal, 1985), and, in view of the comments made previously, it is more likely that phenetic classifications are superior on these grounds. The choice then must be primarily for the phenetic classifications. It is important to note that many of the current approaches to bacterial classification based on ribosomal RNA sequences and related techniques purport to be phylogenetic, but are actually phenetic measures of affinity with molecular sequences as characters. Inferences can then be made about evolutionary patterns but supposed evolutionary pathways are not the basis of the classification. We agree with Jensen (1983) who suggested that what we need are:

1. classifications that reflect what is known about the taxa, and
2. procedures for generating hypotheses about evolutionary relationships.

We shall investigate both these aspects of bacterial taxonomy in the ensuing chapters.

A simplified classification of the bacteria has been included in Appendix 1, and illustrates the enormous complexity of the subject.

REFERENCES

Cain, A. J. (1962). The evolution of taxonomic principles, *Symposium of the Society for General Microbiology*, **12**, 1–3.

Cracraft, J. (1983). The significance of phylogenetic classifications for systematic and evolutionary biology, in *Numerical Taxonomy* (J. Felsenstein, Ed.), pp. 1–17. Springer-Verlag, Berlin, New York and Tokyo.

Gilmour, J. S. L. (1951). The development of taxonomic theory since 1851, *Nature (London)*, **168**, 400–402.

Hennig, W. 1966. *Phylogenetic Systematics*. Universty of Illinois Press, Urbana.

Jensen, R. J. (1983). A practical view of numerical taxonomy or should I be a pheneticist or cladist?, in *Numerical Taxonomy* (J. Felsenstein, Ed.), pp. 53–71. Springer-Verlag, Berlin, New York and Tokyo.

Simpson, G. G. (1961). *Principles of Animal Taxonomy*. Columbia University Press, New York.

Sneath, P. H. A. (1962). Construction of taxonomic groups, *Symposium of the Society for General Microbiology*, **12**, 287–332.

Sneath, P. H. A. (1983). Philosophy and method in biological classification, in *Numerical Taxonomy* (J. Felsenstein, Ed.), pp. 22–37. Springer-Verlag, Berlin, New York and Tokyo.

Sokal, R. R. (1985). The principles of numerical taxonomy 25 years later, in *Computer-Assisted Bacterial Systematics*, (M. Goodfellow, D. Jones and F. G. Priest, Eds.), pp. 1–20. Academic Press, London.

Wiley, J. E. O. (1981). *Phylogenetics: The Theory and Practice of Phylogenetic Systematics*. Wiley, New York.

Williams, J. (1985). Cladistics and the evolution of proteins, in *Computer-Assisted Bacterial Systematics*, (M. Goodfellow, D. Jones and F. G. Priest, Eds.), pp. 61–90. Academic Press, London.

2 Numerical taxonomy

2.1 INTRODUCTION

The early part of this century witnessed dramatic advances in bio-chemistry, genetics and other aspects of biology, but taxonomy plodded on in a haphazard way and contributed little to biology as a whole. Until the late 1950s, bacterial taxa were still recognized on the basis of a few 'key' features, and heterogeneous ill-defined genera were commonplace. Identification of unknown isolates was almost impossible, with the exception of a comparatively few medically important species. With the advent of the computer age, and thus the ability to manipulate large amounts of data rapidly, a development occurred that revolutionized the approach to bacterial taxonomy. Sneath introduced the concept of numerical taxonomy in 1957, and many hundreds of publications have since been devoted to the topic (see Sneath and Sokal, 1973).

Numerical taxonomy, also referred to as NT, Adansonian taxonomy (after the eighteenth century French botanist Michael Adanson), computer taxonomy, numerical phenetic analysis, taxometrics and taxonometrics, is defined by Sneath and Sokal (1973) as: 'the grouping by numerical methods of taxonomic units into taxa on the basis of their characteristics'. Following the basic principles of Adanson, numerical taxonomy necessitates studying as many aspects (traits or characters) of the biology of organisms (referred to as operational taxonomic units; OTUs) as possible. This generates a mass of information on the OTUs under study. However, a key feature of numerical taxonomic methods is that, *a priori*, all the characters have equal importance or weight, a concept that has been difficult for many conventional microbiologists to accept. Indeed this aspect of numerical taxonomy has undoubtedly led to some of the fiercest arguments. Classical taxonomists insist that some tests are more important than others for defining taxa, and that these should be used to establish the classification. Conversely, numerical taxonomists insist that all characters should have equal weight in the construction of classifications. Once the taxa are defined, however, useful differential characters may be weighted (*a posteriori*) for identification purposes.

It is an important premise that taxa should be defined on the basis of overall similarity, and not created simply as a result of taxonomic intuition. This means that taxa are not established solely by the presence (or absence) of certain pre-determined 'essential' characters among groups of OTUs. Thus individual characters, such as the presence of yellow pigmented colonies, are not sufficient to justify the inclusion of OTUs within taxa.

A flow diagram of the stages involved in numerical taxonomy has been included as Fig. 2.1. These stages include strain and test selection, coding of data and their entry into the computer, data analyses, and interpretation of results. The various stages will be considered separately.

2.2. STRAIN SELECTION

With the availability of statistical programme packages, such as CLUSTAN (Wishart, 1978), and powerful computers, numerical taxon-

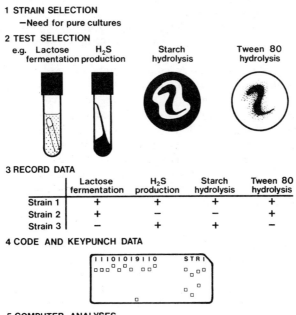

Fig. 2.1 Stages in numerical taxonomy analyses of bacteria.

omy procedures may be applied to large numbers of bacterial strains. Some computers will cope with upward of 600 OTUs in a single set of analyses. However larger data sets may require to be divided, but overall comparisons may be made by means of inter- and intra- group analyses. Although the total number of OTUs may be quite small, as a general guideline data sets should contain information on at least 60 strains. Smaller numbers of OTUs have been used, but apart from studies of very specialized groups or demonstration purposes, it is questionable whether computer-based numerical taxonomy procedures were justified in such cases.

Essentially, studies may be categorized as those that are 'broadly' or 'narrowly' based (restricted studies). The former seeks to study large groups of organisms with a view to defining taxa. These may comprise named strains, such as representatives of bacterial families, e.g., Enterobacteriaceae, or unnamed isolates such as those recovered from ecological studies. Of course, great care is needed in the initial selection of isolates otherwise the outcome of the investigation may be meaningless. These sets of OTUs should include some strains of known identity for comparative purposes. These reference strains serve as markers, and ease the ultimate identification of unknown organisms. It is advantageous to include at least two reference strains of each species, as this reduces the possibility of mistakes due to contamination or mis-labelling, which would negate their comparative value. With this careful attention to detail, numerical taxonomy methods have been used successfully to examine representative isolates from a diverse array of natural habitats, including soil, leaf surfaces and water, and to rationalize large heterogeneous taxa, such as *Streptomyces*. Examples of narrowly based or restricted studies include the investigation of taxonomic relationships and validity between groups of related organisms, for example *Staphylococcus*. These may explore intra- and inter-generic relationships. In most of these cases, named cultures will be used; and it is recommended that the *bona fide* type strains where possible of each species within the area of the study should be used. These reference organisms may be obtained from culture collections such as the American Type Culture Collection (USA) or the National Collection of Type Cultures (UK).

Once the organisms have been selected, it is essential to confirm purity. Good classifications will not result from use of contaminated cultures. Moreover, it is sound policy to maintain separate reserve and working stocks to protect against accidental loss or contamination. The purity of all cultures should be regularly checked during the course of study. For this, the examination of colonial morphology and Gram-stained smears should suffice.

2.3 TEST SELECTION

There are no hard and fast rules governing the choice of tests (characters) to be adopted. The possibilities may appear to be virtually endless, although many characters reflect subtle variations upon central themes. Essentially, the choice may include representatives of the 'classical' phenotypic tests, such as indole production and the ability to ferment lactose, and the 'newer' methods, encompassing chemotaxonomy, for example presence or absence of specific menaquinones (see Chapter 3), or molecular features including DNA homology values (Table 2.1). On balance, it would appear that most numerical taxonomy studies to date have incorporated phenotypic rather than chemotaxonomic traits. The

Table 2.1 Characters used in numerical taxonomy of bacteria

Category of test	Test
Colonial morphology	Presence of non-diffusible or diffusible, fluorescent, non-fluorescent or luminous pigments. Colony size and shape (presence of spreading colonies).
Micromorphology	Gram staining and acid fast staining reactions. Presence of cocci, bacilli, mycelia, sheaths. Cell size. Attachment structures. Presence of intracellular granules. Motility (polar or peritrichous flagella, or gliding). Evidence of di- or pleomorphism Presence of spores.
Growth characteristics	Presence of ring or pellicle, or turbid or flocculent growth in broth. Aerobiosis or anaerobiosis. Growth in 0%–10% (w/v) sodium chloride. Special requirements for growth, eg amino acids, vitamins or metal ions.
Biochemistry	Fermentative or oxidative metabolism of glucose. Presence of catalase, oxidase, and other enzymes. Ability to degrade complex molecules, e.g., starch and tributyrin. Acid production from carbohydrates. Production of indole and H_2S. Methyl red test. Nitrate reduction. Voges Proskauer reaction.
Inhibitory tests	Presence or absence of growth in the presence of antibiotics, dyes and other inhibitory compounds, e.g., potassium cyanide.

Category of test	Test
Utilization of compounds as the sole source of carbon for energy and growth	Presence or absence of growth in the presence of carbon containing compounds, e.g., alanine, fructose, maltose and sodium citrate.
Serology	Presence or absence of agglutination reactions to specific antisera.
Chemotaxonomy	Presence or absence of sub-cellular components, e.g., mycolic acids and menaquinones.
Molecular genetics	Presence of specified guanine plus cytosine (G + C) ratios of the DNA.
Phage typing	Presence of specified bacteriophage typing patterns

ideal situation would be to use characters representing the expression of single genes or operons that are not subjected to environmental changes. Such characters should be stable, and therefore enable reliable natural classifications to result. Tests, based on single properties, are referred to as unit characters, and by definition comprise taxonomic characters of two or more states, which cannot be subdivided logically, except for changes in the method of coding. Needless to say unit characters vary in the amount of genetic information they represent. The presence of an endospore and the utilization of sucrose are both unit characters, but the former represents some 50 operons whereas the latter just one. With the current awareness of microbial molecular biology, it is impossible to subdivide sporulation into the constituent characters and, for practical purposes it is considered as a single or unit character. It is over this point that critics of numerical taxonomy and the *a priori* equal weighting of characters are generally most vociferous, but the practical concept of the unit character is well established and to many the best solution.

In ideal circumstances, the testing regime should comprise a random assortment of all the possible characters which could be studied. This would preclude the temptation of *a priori* weighting of the characters through choice of specific features. However in practice, it is satisfactory to use a battery of tests representing as many facets of the biology of the organisms as possible. An ideal list will include colony and micromorphology data, growth characteristics, biochemical tests, effect of potentially inhibitory agents, utilization of compounds as the sole source of carbon for energy and growth, and serological, chemotaxonomic and molecular genetic information (Table 2.1). It is a sensible precaution to use approximately the same number of characters from each of the general categories mentioned above, as this will minimize influencing the outcome of the exercise in favour of certain facets, for example classifications based heavily on colonial and micromorphological features. When surveying the tests to be included, it is prudent to select those that are simple and inexpensive in both time and money because a large number of tests will be carried out and complicated time-consuming chemical analyses would not be appropriate. Redundant tests should not be used. These include characters that are wholly positive or negative for the OTUs under study, and would, therefore, have no discriminatory value. As an extreme example, a test scoring the presence or absence of DNA in a bacterial cell would contribute nothing meaningful to the classification. Unfortunately uniformly negative or positive tests are often not known until all the tests have been completed! Poorly defined or poorly reproducible tests should also be avoided since ambiguous or wrong answers would result. A fluctuation in recording poorly defined tests is likely to occur, particularly if a large number of OTUs are under study. An example might concern an over-ambitious attempt to define colony colour in terms of slight differences between white, off-white, cream, pale yellow and yellow. The interpretation of such slight colour variations is likely to vary considerably from the start to finish of any testing regime and would introduce an error component into the study.

The optimum number of characters to be studied for each OTU would appear to lie between 100 and 200. Greater numbers have been used by some workers, but the increased information content of the finished study rarely justifies such enthusiasm. Essentially classifications are not substantially improved by the huge numbers of characters, and the optimum number seems to be 100–150. In contrast, the effective minimum is approximately 50–60 characters. The obvious danger from smaller batteries of tests is that each of the characters would exert a disproportionately large influence on the analyses, particularly if any errors occurred. However the problem of test error and, for that matter, test reproducibility will be considered in the next section. As far as possible, each character should be recorded for each OTU because incomplete data may influence the outcome of the computation.

Careful consideration needs to be given to the exact methods employed. Essentially there needs to be strictly standardized conditions of medium formulation, inoculation procedures, and incubation times and temperatures. Any special condition required by one OTU *must* be applied to all OTUs under study. For example if one OTU requires 1.0% (^{w}/v) sodium choloride and/or 0.1% (^{w}/v) L-cysteine hydrochloride for growth, then all strains should be exposed to the same concentrations of these compounds. It is desirable that all media should be inoculated with a standard quantity of cells, preferably in the logarithmic phase of growth. Moreover, all strains should be incubated at a common (optimum) temperature for the same period of time. This necessitates the use of sufficient incubation periods to permit the growth of the slowest organism. Where possible, control strains should be incorporated into the study, to check on the sensitivity of media and reagents and their reactions.

2.3.1. Test error/reproducibility

Although many scientists may regard their work as an utopian example of perfection, experience suggests otherwise. It is unfortunately very easy for mistakes to creep into the data, whether reflecting errors in recording or uncertainty in the correct interpretation of test results. It is noteworthy that freshly isolated environmental strains are particularly capable of losing metabolic activities, such as synthesis or degradation of complex molecules upon repeated sub-culturing in the laboratory. Perhaps this reflects the loss of plasmid DNA or other more subtle changes in the bacterial physiology. Problems abound in the 'correct' interpretation of so-called borderline reactions; is the result weakly positive or negative? As stated previously in studies of large numbers of OTUs, variations in recording occur during the course of the study. This problem of test error/reproducibility in numerical taxonomy was voiced initially by Professor P.H.A. Sneath (e.g. Sneath and Johnson, 1972). The outcome was a strong recommendation that an estimate of test error should form an integral part of any numerical taxonomy study. Essentially this involves the use of randomly picked duplicate cultures, the identity of which should be concealed to reduce experimenter bias. These cultures, amounting to approximately 10% of the total number of OTUs should be included with the collection of strains and examined throughout the testing regime. On completion, the data from the duplicate cultures are used to estimate test error. The variances of individual tests between replicate organisms (S_i^2) may be calculated from the equation:

$$S_i^2 = \frac{n}{2t} \qquad \text{(Equation No. 15; Sneath and Johnson, 1972)}$$

Here, n corresponds to the number of OTUs with discrepancies in the test, and t is the total number of strains. This may be illustrated by examination of the following specimen data set in Table 2.2.

Thus the individual test variance (S_i^2) among the duplicate strains for

Table 2.2 Specimen data set

| | OTU | Test | | |
		A	B	C
Original	1	+	+	−
Duplicate culture	1	+	−	−
Original	2	+	+	+
Duplicate culture	2	−	+	+
Original	3	+	+	+
Duplicate culture	3	+	−	+

tests A, B and C are 0.16, 0.33 and 0, respectively. The individual test variances may be averaged to provide a global estimate of error. The formula is:

$$S^2 = \frac{1}{N}(S_A^2 + S_B^2 \ldots S_N^2),$$

where N equals the total number of tests, and S_A^2, S_B^2 aand S_N^2 correspond to the individual test variances for tests A, B and up to N, respectively. Continuing with the above example:

$$S^2 = \frac{1}{3}(0.16 + 0.33 + 0) = 0.161$$

The probablility of error for an individual test (P_i) is:

$$P_i = \frac{1}{2}[1 - \sqrt{(1-4\,S_i^2)}] \quad \text{(Equation No.4; Sneath and Johnson, 1972)}$$

The pooled variance (S^2) may therefore be used to determine the average probability of an erroneous test, ie test error among the data. For the above example:

$$P_A = \frac{1}{2}[1 - \sqrt{(1-4 \times 0.16)}] = 0.2 = 20\%$$
$$P_B = \frac{1}{2}[1 - \sqrt{(1-4 \times 0.33)}] = 0.22 = 22\%$$
$$P_C = \frac{1}{2}[1 - \sqrt{(1-4 \times 0)}] = 0$$

where:

$$P = \frac{1}{2}[1 - \sqrt{(1 - 4 \times 0.161)}] = 0.29 = 29\%$$

In cases where $P = > 10\%$ (as in the above example!), there may be serious problems with the reliability of the numerical taxonomy study, and an erroneous measurement of similarity between the OTUs may result. It is a wise precaution to reject individual tests which show error of > 10–15%. Ironically, some of the classical bacteriological tests such as nitrate reduction, oxidase, gelatinase and urease production appear to enter this category. However, the evidence points to poorer test reproducibility at the inter- rather than intra-laboratory level. For numerical taxonomy, it should suffice to use large numbers of characters, thereby reducing considerably any detrimental effects caused by a small number of poorly reproducible rogue tests.

2.4   DATA CODING

With completion of the testing regime and the production of neat and comprehensive data books, the next stage is to code the information in a format suitable for computation. A table, referred to as an $n \times t$ table, is produced, which lists the test results for all the OTUs in the investigation. Unit characters, which exist in either of two states, i.e., presence or absence, are coded numerically as '1' and '0' corresponding to '+' and '−' respectively (Table 2.3). Some programme packages can accept missing or non-comparative data, which are coded usually as either '3' or '9'. It must be emphasized that there is likely to be local variations in the required data format. Therefore, enquiries should be made at the local computer centre to ascertain the precise requirement.

Table 2.3   Examples of characters and character states used in the numerical taxonomy of bacteria

Character	Character state
Colony coloration	White/off white/yellow/orange/red/purple
Diffusible fluorescent pigment	Presence/absence
Gram-staining reaction	Gram-positive/Gram-negative/Gram-variable
Micromorphology	Rods/cocci/mycelia
Motility	Presence/absence
Catalase production	Presence/absence
Nitrate reduction	Presence/absence
Degradation of starch	Presence/absence
Growth on 7% (^{w}/v) sodium chloride	Presence/absence
Utilization of individual compounds as the sole source of carbon for energy and growth	Presence/absence

Qualitative data may be reduced to two or more multistate characters (Table 2.4). In the example of colony pigmentation, four possible character states have been defined. Of course, it could be argued that there has been *a priori* weighting of pigmentation by splitting it into four characters. Thus a colony that is purple is not, of course, orange or red. This gives these different colony types as much in common as they have distinguishing them. This is a sensible criticism, and such unintentional weighting may be reduced by not computing the redundant characters for each OTU.

Data, for which an OTU is assessed quantitatively, may be scored as a series of unit characters or directly as numerical values. Production of amylase as zones of starch hydrolysis or susceptibility to antibiotics determined as zones of clearing around discs might come into this category. In these cases, the number of columns used is one less than the total number of incremental points (see Table 2.5). Again, it could be

Table 2.4 Qualitative coding of characters

Strain number	Character			
	Presence of: red colonies	orange colonies	purple colonies	cream colonies
1	1	0	0 (nc)	0 (nc)
2	0	1	0	0 (nc)
3	0 (nc)	0	1	0
4	0 (nc)	0 (nc)	0	1

nc = not computed.

Table 2.5 Examples of Quantitative coding of characters

Strain number	Characters		
	Slight susceptibility to streptomycin	Moderate susceptibility to streptomycin	Highly susceptible to streptomycin
1	0	0	0
2	1	0	0
3	1	1	0
4	1	1	1

nc = not computed.

argued that there has been *a priori* weighting of the characters, but, in fact, additive coding retains the dimension of the character and only applies excessive weight if the character has been divided into too many unit characters. An example of a quantitative character coded directly as a numerical value would be the colony diameter measured in mm.

The data may now be entered into the computer either directly via remote terminal or by key-punching data cards. Verification of the coded data is a thankless task which serves to check for mistakes. Assuming that all is well, the data are now ready for computer analyses.

2.5   COMPUTER ANALYSES

Since the initial developmental work, in the 1950s, computational procedures have evolved with ever-increasing sophistication. Jargon has become widespread, and it is understandable for the novice taxonomist to be frightened by the esoteric statistical language. For the present, it suffices to emphasize that the initial role of the computer is to compare the data for each OTU with every other OTU. This enables the level of similarity or conversely dissimilarity (known also as distance) to be calculated between each and every OTU. The most widely used methods in bacterial taxonomy calculate similarity values. Therefore, these techniques will be discussed initially.

2.5.1   Calculation of resemblance

Only a few of the many possible methods for measuring similarity values find routine use in bacteriology. The formulae of some potentially useful methods, referred to as coefficients, are shown in Table 2.6. Quite simply, these coefficients measure the relationships between pairs of OTUs and have been devised with diverse disciplines in mind, including botany, paleontology and zoology. Nevertheless, they serve essentially the same purpose. From Table 2.6, it will be observed that these formulae are based on a group of symbols; a, b, c and d. These correspond to the number of positive (a) and negative (d) matches, and the number of dissimilar results (b, c) between a pair of OTUs. This relationship may be represented diagrammatically as:

		Results for OTU1	
		+	−
Results for OTU2	+	a	b
	−	c	d

Table 2.6 Some coefficients with useful discriminating value for microbiology (After Austin and Colwell, 1977)

Coefficient	Abbreviation	Formula
Simple matching	S_{SM}	$\dfrac{a + d}{a + b + c + d}$
Jaccard	S_J	$\dfrac{a}{a + b + c}$
Dice	S_D	$\dfrac{2a}{2a + b + c}$
Hamann	S_H	$\dfrac{(a + d) - (b + c)}{a + b + c + d}$
Kulczynski	S_{K2}	$\frac{1}{2}[\{a/(a + b)\} + \{a/(a + c)\}]$
Ochiai	S_O	$\dfrac{a}{\sqrt{[(a + b)(a + c)]}}$
Pattern difference	D_P	$\dfrac{2\sqrt{(bc)}}{a + b + c + d}$
Rogers and Tanimoto	S_{RT}	$\dfrac{a + d}{a + d + 2(b + c)}$
Total difference	D_T	$\dfrac{b + c}{a + b + c + d}$
Unnamed	S_{UN1}	$\dfrac{2(a + d)}{a + b + c + d}$
	S_{UN4}	$\frac{1}{4}[\{a/(a + b)\} + \{a/(a + c)\}\{d/(b + d)\} + (d/a)]$
Angular transformation of $S_{SM} \times 2/\pi$	$\text{Sin}^{-1}(S_{SM})$	$0.637 \times \arcsin\left(\sqrt{\dfrac{a + d}{a + b + c + d}}\right)$

a and d correspond to the number of positive and negative matches, respectively.
b and c represent the number of non-matching characters between pairs of OTUs.

To illustrate this calculation, it is necessary to examine a sample data set:

OTU	Results of individual (unnamed) tests:
1	+ + + − − − + − + − + + − + − + − + − +
2	+ − + + − − − + − + + − + − − + + + − +

For this 20 test data set, the number of positive matches, (the a component) between the pair of OTUs is 6. Similarly, the number of negative matches (the d component) is 4. The number of tests for which OTU1 is negative but which OTU2 is positive, (the b component) is 5, and the converse of b (the c component) is 5. The summation of $a + b + c + d$, is the total number of tests, which in this example is 20. One of the most widely used coefficients is the simple matching coefficient (S_{SM}), which measures positive matches (a) and negative matches (d) as a proportion of the total number of characters $(a + b + c + d)$. Thus, $S_{SM} = (a + d)/(a + b + c + d) = 0.5$. The scale of values with this coefficient will range from 'zero', for total dissimilarity, to 'one' for total similarity between pairs of OTUs.

A second coefficient, which finds widespread use is the Jaccard coefficient (S_j), which takes its name from the classic study of Jaccard in 1908. This coefficient discounts negative matches. Thus, $S_J = a/(a + b + c) = 0.38$.

The range of values also spreads from zero to unity and it is straightforward to convert the values into percentages. The S_J and S_{SM} coefficients have been used in most publications concerning numerical taxonomy studies of bacteria. It will be readily apparent that differences in similarity values between OTUs will result according to the nature of the coefficient used. This begs the question about which is the correct method to be used, since the choice of coefficient will affect the outcome of classification. This dilemma will be addressed later.

Another coefficient, which has been used occasionally in bacteriology, is Gower's coefficient (S_G). This is a weighted average of all similarity values between pairs of OTUs and is suitable for use with binary, quantitative and qualitative data. For binary data S_G is equivalent to the S_J coefficient, whereas for qualitative characters of two or more states S_G is identical to the S_{SM} coefficient. However for quantitative data:

$$S_G = \left[1 - \frac{\text{(the value of OTU1 for a given character minus the value of OTU2 for the same character)}}{\text{the range of values for the character}} \right]$$

A fairly complex example may be illustrated by use of a sample data set (Table 2.7).

Table 2.7 Sample data set

OTU	Motility	Colony diameter (mm)	Guanine plus cytosine (G+C) ratios of the DNA (mol %)	Colony colour
1	+	2.5	48	Orange
2	+	2.3	62	Yellow
3	−	1.8	43	Cream
4	−	1.5	42	Cream

The range of values for colony diameter, i.e., 1.5 to 2.5 mm, is 1.0. For the G + C ratio, the range, i.e., 42 to 62 mol %, is 20. Dealing with the calculation for S_G of OTU1 and OTU2, it may be determined that:

$$S_G = \frac{\overset{i}{1} + \overset{ii}{(1 - 0.2/1.0)} + \overset{iii}{(1 - 14/20)} + \overset{iv}{0}}{1 + 1 + 1 + 1} = 0.525$$

Here 'i', 'ii', 'iii' and 'iv' refer to the presence of motility, colony diameter, G + C value, and colony colour, respectively. For 'iv', the zero score on the top line refers to the non-match of colour between the OTUs. The $1 + 1 + 1 + 1$ score for the bottom line refers to the presence of four characters. Use of S_G for OTU's 3 and 4 shows that:

$$S_G = \frac{\overset{i}{0} + \overset{ii}{(1 - 0.3/1.0)} + \overset{iii}{(1 - 1/20)} + \overset{iv}{1}}{0 + 1 + 1 + 1} = 0.883$$

It may now be obvious that the 'i' component has been scored as zero because of the absence of motility. However, colour 'iv' is scored as unity because of the match, ie both OTUs produce cream colonies.

Obviously, there is a requirement for many arithmetic calculations to compare data for all the OTUs in the study. In fact, the total number of calculations between OTUs is $N = \frac{1}{2} n (n-1)$, where n is the total number of OTUs. For groups of 50, 100 and 200 OTUs, it would be necessary to carry out 1225, 4950 and 19 000 calculations, respectively. Thus there is an exponential increase in the number of calculations commensurate with a linear increase in the number of OTUs.

2.5.2 Distance coefficients

Both novice and experienced numerical taxonomists may express reservation about the terminology referring to dissimilarity measurements. These measurements, which are also referred to as metrics and distance (D), are the converse of, and are complementary to, methods of assessing similarity. In this section, discussion will centre around Euclidean distance, which has been used in numerical taxonomy studies of bacteria. An excellent description of the dissimilarity measurement is to be found in Dunn and Everitt (1982). It is important to emphasize that this measurement satisfies the conditions of Pythagoras's theorem, which concludes that the squared value of the hypotenuse of a right-angled triangle equals the sum of the squared values of the other two sides. In brief, Euclidean distance measures the distance between OTUs in terms of their co-ordinates on right-angled (also referred to as Cartesian) axes. The values extend from zero to an infinitely large amount depending upon the number and size of the differences between OTUs. Therefore, these values may not be transposed to percentages. As Euclidean distance is determined from Pythagoras's theorem, D^2 is calculated, which is itself used as a measurement of dissimilarity. The advantage of using D^2 is that it avoids the necessity of calculating square roots. In its simplest form for binary data, D corresponds to $\sqrt{(1 - S_{SM})}$ and gives the diagonal distance between two OTUs. However in this case, use of D^2 is beneficial insofar as the dissimilarity between OTUs corresponds to the number of binary characters by which they differ, i.e., the b and c component of the previously mentioned notation. This assumes that missing data are not present in the data matrix. Therefore in these circumstances $D^2 = (b + c)/(a + b + c + d)$, whereas $D = \sqrt{(b + c)/(a + b + c + d)}$. For quantitative multistate characters, the Euclidean distance between a pair of OTUs is defined as:

$D = \sqrt{}$(the value of the first OTU for a given character minus the value of the second OTU for the same character)^2

This calculation continues by adding similar information for the second character, continuing until all of the characters have been so considered. Euclidean distance may be expressed as (see, for example, Clifford and Stephenson, 1975):

$$D = \left[\sum_{I}^{n} (\chi_{OTUA} - \chi_{OTUB})^2 \right]^{1/2}$$

where 'sigma' is the summation sign for all the characters from the first, i.e. I to the last, i.e., Nth; and χ_{OTUA} and χ_{OTUB} correspond to the value of OTUA for character χ and the value of OTUB for the same character, respectively. The determination of Euclidean distance for quantitative data may be illustrated, using the following example:

OTU	Length of cells (μm)	G + ratio of DNA	Greatest quality of sodium chloride permitting growth (%)
1	2.0	42	1.5
2	2.5	45	2.0
3	3.0	59	6.0

For OTU1 compared to OTU2:

$$D^2 = (2.0 - 2.5)^2 + (42 - 45)^2 + (1.5 - 2.0)^2 = 9.5$$

$$D = 3.08$$

For OTU1 measured against OTU3:

$$D^2 = (2.0 - 3.0)^2 + (42 - 59)^2 + (1.5 - 6.0)^2 = 310.25$$

$$D = 17.61$$

Similarly for OTU2 assessed against OTU3:

$$D^2 = (2.5 - 3.0)^2 + (42 - 59)^2 + (2.0 - 6.0)^2 = 212.25$$

$$D = 14.56$$

From this example, it is apparent that OTU1 is closer to OTU2 than to OTU3.

The outcome from all the similarity and distance analyses is an unsorted (similarity or dissimilarity) matrix that records the similarity/distance between each of the OTUs. In practice, this is a triangular array, and approximates to mileage charts showing distances between towns.

2.5.3 Vigour and pattern statistics

A further significant development by Sneath in 1968 concerned the effect on bacterial taxonomy of vigour and pattern elements. It was reasoned that the total difference between OTUs reflected two components, namely pattern differences and vigour differences, which may be equated with shape and size difference, respectively, in animals or plants. The object of Sneath's work was to reduce the apparent dissimilarity between OTUs that could be attributed to differences in growth rate, i.e., the vigour difference, because slow-growing strains may be recorded as negative in a test, such as acid production from sugars, not because they lack the necessary catabolic pathway but because they do not grow fast enough for a positive reaction to be recorded within the specified time period. This coefficient is, therefore, useful when strains of widely different metabolic activity are being studied. The vigour of an organism was defined as the proportion of its characters that showed a positive response in the tests examined, and the difference in vigour (D_v) is simply the difference in this value. According to Sneath, there is a relationship between vigour and pattern, as follows:

(Total difference; $D_T)^2$ = (Vigour difference; $D_v)^2$ + (Pattern difference; $D_p)^2$

i.e.

$$D_T^2 = D_v^2 + D_p^2$$

In essence: $D_T = 1 - S_{SM}$

Using the previously described symbols of a, b, c and d, it should be apparent that the total difference (D_T) between a pair of OTUs is:

$$\frac{b + c}{a + b + c + d}$$

Vigour, being considered as the proportion of positive matches, is regarded as:

$$\frac{a + c}{a + b + c + d}$$

for OTU1 and:

$$\frac{a + b}{a + b + c + d}$$

for OTU2. The vigour difference between OTU1 and OTU2 is the difference in these values, i.e.:

$$\left(\frac{a + c}{a + b + c + d}\right) - \left(\frac{a + b}{a + b + c + d}\right) = \frac{c - b}{a + b + c + d} = D_v$$

It is apparent that D_v may be positive or negative, depending upon the relative size of c to b.

The pattern difference (D_p) is obtained from the equation:

$D_p{}^2 = D_T{}^2 - D_v{}^2$. Thus:

$$D_p = \frac{2 \sqrt{(bc)}}{a + b + c + d}$$

D_p ranges in value from zero for identical pairs of OTUs to unity for completely dissimilar pairs of OTUs.

Although both D_v and D_p were initially devised for binary characters, the concept has been extended for multistate quantitative characters. Vigour and pattern statistics have been well received by bacterial numerical taxonomists, as attested by the widespread use. For an example of these analyses see Goodfellow *et al.* (1976).

2.6 DETERMINATION OF TAXONOMIC STRUCTURE

2.6.1 Hierarchical clustering

The ordering of OTUs into groups of high overall phenetic similarity is generally achieved by means of one of several commonly used hierarchical clustering methods. Popular procedures include single- and average-linkage clustering. In addition, there is limited interest among bacterial taxonomists for complete-linkage, centroid sorting and Ward's method. The relative value of these methods will be considered later.

In the case of the linkage methods, the algorithm starts by searching the data matrix for the highest value between pairs of OTUs. The codes for, and the similarity between, these OTUs are then listed. This pair with the highest similarity value is treated as a single OTU. The computer then computes the next highest pair, which may be between two other OTUs or the previous pair and another OTU. The cycle proceeds until all the OTUs are included in clusters ($T - 1$ cycles for T OTUs).

The clustering techniques differ in the definition of similarity between OTUs and groups, and between groups. In single linkage (nearest neighbour) clustering, OTUs join groups at the highest similarity between the OTU and any *one member* of the group with no account being taken of other members of the group. Thus in the example shown in Fig. 2.2, OTU5 joins cluster 2, 3, 1, 4 at 50% similarity because it shares this with OTU4 despite it possessing a lower similarity with the other members of the cluster. Similarly clusters 2,3 and 1,4 coalesce at 70% through the relatedness of 1 and 3 (70%) and/or 2 and 4 (70%) despite the lower levels shown by the pairs 1, 2 and 3, 4.

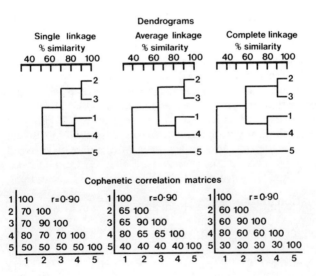

Data (similarity) matrix

OTU					
1	100				
2	60	100			
3	70	90	100		
4	80	70	60	100	
5	40	30	40	50	100
	1	2	3	4	5

Results of clustering

% Similarity	Single linkage	Average linkage	Complete linkage
100	1 2 3 4 5	1 2 3 4 5	1 2 3 4 5
90	2,3 1 4 5	2,3 1 4 5	2,3 1 4 5
80	2,3 1,4 5	2,3 1,4 5	2,3 1,4 5
70	2,3,1,4 5	2,3 1,4 5	2,3 1,4 5
60	2,3,1,4 5	2,3,1,4 5	2,3,1,4 5
50	2,3,1,4,5	2,3,1,4 5	2,3,1,4 5
40	2,3,1,4,5	2,3,1,4,5	2,3,1,4 5
30	2,3,1,4,5	2,3,1,4,5	2,3,1,4,5

Dendrograms

Single linkage   Average linkage   Complete linkage

% similarity

40 60 80 100

Cophenetic correlation matrices

1	100	r=0·90			
2	70	100			
3	70	90	100		
4	80	70	70	100	
5	50	50	50	50	100
	1	2	3	4	5

1	100	r=0·90			
2	65	100			
3	65	90	100		
4	80	65	65	100	
5	40	40	40	40	100
	1	2	3	4	5

1	100	r=0·90			
2	60	100			
3	60	90	100		
4	80	60	60	100	
5	30	30	30	30	100
	1	2	3	4	5

Fig. 2.2 Process of single, complete and average linkage cluster analysis. A comma between two strain numbers denotes that the two strains are included in a cluster (see text for details).

Complete linkage clustering (furthest neighbour) incorporates OTUs in clusters or combines clusters at the lowest similarity level between the OTU and *any member* of the cluster. Thus in the example in Fig 2.2, OTU5 joins the cluster at 30% similarity, the value for the pair 5 and 2. Similarly the clusters 2, 3 and 1, 4 coalesce at 60%, the values for the pairs 3, 4 and 1, 2, respectively.

Average-linkage, or to give it its full name of unweighted pair group method with arithmetic averages (UPGMA) is a compromise of the single and complete linkage algorithms. In this procedure, an OTU joins a group at the average similarity between the OTU and all members of the group. Again, referring to the example in Fig 2.2, OTU5 joins the cluster 2, 3, 1, 4 at the average similarity which is $40 + 30 + 40 + 50 = 160/4 = 40\%$. Similarly the two groups 2, 3 and 1, 4 fuse at the mean similarity of the group individual relatedness values; $2,1 = 60\%$; $2,4 = 70\%$; $3,1 = 70\%$; $3,4 = 60\%$ which is $60 + 70 + 70 + 60 = 260/4 = 65\%$ Although it takes longer in computer time, it can be shown theoretically that average-linkage cluster analysis gives the most accurate representation of the taxonomic structure.

Centroid sorting/clustering joins pairs of OTUs (or an OTU into a group of OTUs) according to the co-ordinates of their centroids (average value) when considered in multidimensional terms. Groups are joined at the minimum value/distance between the centroids. One major disadvantage of this method is that the identity of small groups of OTUs may be lost upon merging with a large group. In this case, the new centroid position may now fall entirely within the area delineated by the large group. However, it may be necessary to maintain the relative identity of the original smaller group. This problem was resolved by Gower, who described the technique of median sorting. With this, groups are considered to occupy a unit size, for which an unweighted median position is obtained after fusion.

The fifth clustering method, referred to previously, is Ward's method. With this technique, clustering is based on the minimum sum of squares within clusters.

2.6.2 Ordination methods

Apart from cluster analyses, which account for most numerical taxonomy studies of bacteria, OTUs may be ordered on the basis of two- or three-dimensional arrays. These are ordination diagrams (also known as taxonomic maps), which are non hierarchic and constitute the so-called multivariate analyses. With such methods it is possible to view the relationship between OTUs in terms of phenetic (taxonomic) hyperspace. It is important to emphasize that ordination methods (the term was coined by the botanist, Goodall, in 1953) do not divide OTUs into convenient groups, as may be achieved by cluster analyses, although groups may be recognized by careful examination of the output. Instead, ordination methods concern establishing appropriate representations between OTUs in terms of distance and space. Thus these methods are useful for the examination of entities in which the relationship between them represents a continuum, such as probably occurs with bacteria, rather than the more artificial discrete-like 'boxes' highlighted by traditional taxonomy (see Alderson, 1985).

Ordination methods have the potential to be more informative, in terms of relationships between OTUs, than dendrograms. The former usually provide a realistic measure of distance between principle groups, although such distances may be extremely artificial between close neighbours. It is apt to consider the three-dimensional arrangement of planets in space as a parallel to this aspect of bacterial taxonomy. Of course the arrangement will largely depend on the actual position from which the observations are made. This should be borne in mind during consideration of ordination methods.

A method, in regular use, is the principal component analysis, which was first proposed by Pearson in 1901. With this analysis, the relative positions of the OTUs, as measured in terms of distances from each other, are plotted on right-angled (Cartesian) co-ordinates (axes). However it should be emphasized that these axes must be chosen such that the positioning of the OTUs will most appropriately demonstrate their relationships (see Fig. 2.3). A straight line may then be drawn through

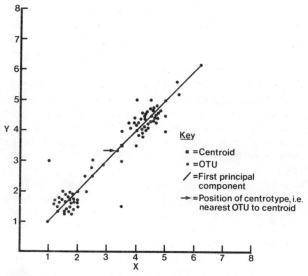

Fig. 2.3 Example of principal components analysis.

the position of the OTUs, passing through the centroid. Incidentally, the nearest OTU to the centroid is the centrotype. Thus, the location of as many OTUs as possible should be on the line, and the rest are in equal numbers above and below the line. Technically, it may be described that the position of the line is such that the summation of perpendicular measurements from the pairs of coordinates for each OTU to the line is the minimum value (Clifford and Stephenson, 1975).

The line between the x and y axes, is shown in Fig. 2.3. This line, along which the maximum number of OTUs is spread, is referred to as the first principal component. A line at right angles, to this first line is the second principal component. To determine how to extract principal components from dissimilarity matrices, the interested reader is referred to the excellent text of Dunn and Everitt (1982). A similar technique is principal coordinate analysis, which enables certain 'important' features of the data to be emphasized.

Two other noteworthy multivariate procedures include canonical variate (known also as multiple discrimination) analysis and canonical correlation analysis. However these analyses use data for OTUs which have been already grouped, and the relationship between the taxa in multi-dimensional space is investigated (see Chapter 6).

2.7 PRESENTATION AND INTERPRETATION OF RESULTS

The comparatively large volume of paper produced as a result of computer taxonomy contains essential facets of information for the construction of a sorted similarity (dissimilarity) matrix and a dendrogram or the multidimensional diagrams from ordination procedures. The matrix is a triangular array of numerical results, which permits the comparison of an OTU with any other listed in the data set (Fig. 2.4a). For the S_{SM} and S_J coefficients, the results are usually printed to three decimal points. Although there may be initial confusion as to the reasons for a triangular and not a square matrix, it is apparent that any further information necessary to complete a square would be a mirror image. Multiplication of these values by 100 gives rise to percentages, and thus a slightly simplified version of the similarity matrix (Fig. 2.4b). Thereafter, numerical values may be converted to shaded diagrams (Fig. 2.4c and 2.4d), in which different intensities of shading are used to represent bands of similarity values. It is good policy for the most and least intense shading to correspond with the highest and lowest similarity values, respectively. In Fig. 2.4(c), values to up to 80% and above 90% have been banded for convenience in groups of ten percentage points. Between 81 and 89%, two divisions for the shading have been used. Interpretation of shaded diagrams is much easier than the numerical values expressed in Fig. 2.4(a) and 2.4(b). The areas of high overall similarity are readily

Modern Bacterial Taxonomy

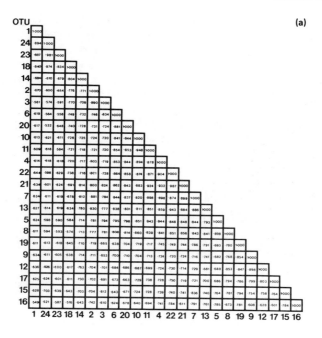

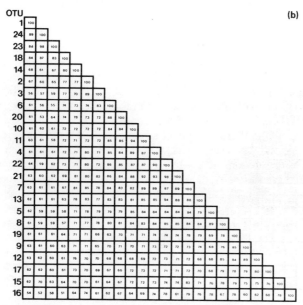

Fig. 2.4 Stages in the production of a sorted similarity matrix. The initial information, to three decimal places of accuracy (a), is converted to percentage values (b), and then to shaded diagrams (c,d). The relatedness between OTUs is

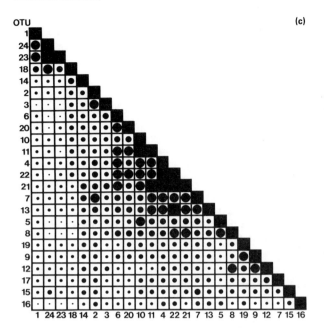

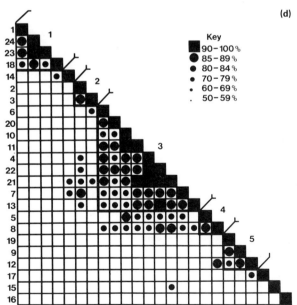

readily apparent by eliminating the lower values from the diagram (d). Numbers 1 to 5 are cluster numbers.

observed, and thus clusters of related OTUs may be defined. It is relevant to note that at least one computer package, e.g., UMDTAXON available at the University of Maryland, is capable of generating simplified similarity matrices. With this system, symbols of the type available on most typewriters are used to represent different levels of similarity. A further refinement involves omitting the low similarity values, in which case areas of high similarity are even more readily observed (Fig. 2.4d). Examples of shaded diagrams of similarity matrices are regularly published in bacteriology journals (e.g., Goodfellow *et al.*, 1976).

Cluster analysis also generates dendrograms, which may be printed directly via graph plotters or constructed manually from salient details (Fig 2.5). Nevertheless, the dendrogram may in turn be simplified by grouping together OTUs clustering together at pre-determined levels of similarity. These groups may be presented as shaded triangles, in which the length of the side beneath the 100% similarity level is proportional to the number of OTUs (Fig. 2.5b).

The diagrams, illustrated by Figs. 2.4 and 2.5 permit groups/clusters of related OTUs to be recognized. These groups, defined on the basis of overall phenetic similarity, are termed phenetic groups or phena. In Fig. 2.4(d) and 2.5(b), these phena have been labelled as '1' to '5'. It is generally accepted that 'species' should be defined at the 80 to 85% similarity (S) level if using S_{SM} – UPGMA generated dendrograms. All strains that

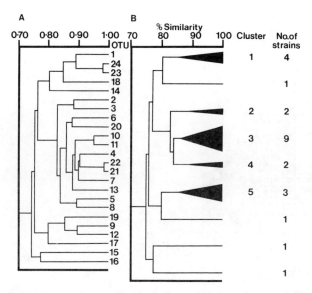

Fig. 2.5 A dendrogram produced by computer-based graphics sub-routine (a) may, in turn, be simplified by reducing clusters to triangular areas of shading (b).

coalesce above this level may be considered as presumptive members of a species; confirmation by chemotaxonomic markers is often desirable, however. Similarly, a generic boundary can sometimes be drawn across a dendrogram at 60–65% similarity, but again this needs to be substantiated by chemotaxonomic evidence. It should be emphasized that these similarity levels are somewhat arbitrary, having been delineated as a result of (numerous) past experiences of comparing the output of numerical analyses with those of conventional classifications. Of course the wisdom of such action is debatable, and, to some extent, undermines the critical basis of numerical taxonomy. However, the existing evidence, particularly for the Enterobacteriaceae, would support these similarity levels for the definition of species and genera. For other groups, such as the *Flavobacterium – Cytophaga – Flexibacter* group of bacteria, it is difficult to make such generalizations. Yet, the dendrogram can provide a clear and accurate picture of the taxonomic structure of a group of organisms. A working example has been included in Fig. 2.6.

Examples of output from ordination methods have been illustrated in Fig. 2.7. As with cluster analysis, it is possible to delineate groups of

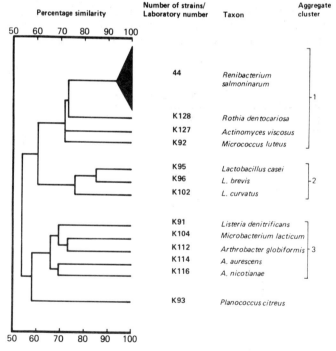

Fig. 2.6 A dendrogram showing the relationship between *Renibacterium salmoninarum* and other Gram-positive organisms based on the S_{SM} coefficient and unweighted average linkage and clustering (from Goodfellow *et al.* 1985).

OTUs. These may be differentiated from each other by means of various shadings or shapes (see Fig. 2.7) or by enclosing groups of OTUs by circles. One of the main advantages over cluster analyses is that ordination methods permit relationships to be revealed which could be masked by dendrograms.

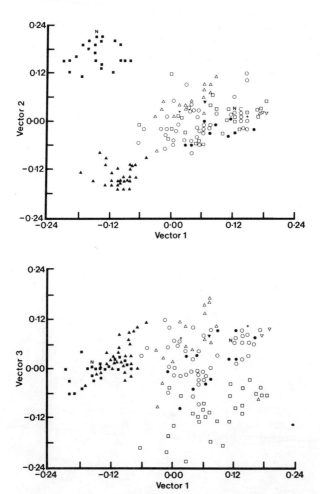

Fig. 2.7 Plot of the first 3 vectors, accounting for 38% of the total variation, from a principal coordinate analysis of some *Bacillus* strains. *Bacillus anthracis,* ■ ; *B. cereus,* o; *B. cereus* strains from diarrhoeal-type food poisoning outbreaks, ● ; *B. cereus* strains from emetic-type food poisoning outbreaks, and serotypes 1, 3, 5 and 8, ▲ ; *B. cereus* var. *albolactis,* ◇ ; *B. cereus* var. *fluorescens,* ▽ ; *B. cereus* var. *mycoides,* △ ; *B. praussnitzi,* + ; *B. thuringiensis,* □ ; *B. filicolonicus,* ▼ ; neotype strain, N. (From Logan and Berkeley, 1981.)

Thus, the outcome of any numerical taxonomy study will reflect the nature of the methods used. Perhaps a legitimate criticism is the lack of standardization in methods, starting with the intitial choice of OTUs and the numbers and types of characters to be used. Thereafter, the wide range of computer-based analyses may daunt any taxonomist. The effect of similarity coefficients on numerical taxonomy studies of bacteria has already been addressed (see Austin and Colwell, 1977). Such differences between the use of the S_I, S_{SM}, S_{K2} and S_{UN4} coefficients (see Table 2.6 for the formulae) for Enterobacteriaceae representatives have been highlighted in Fig. 2.8. It would be pertinent to enquire which (if any) is correct. Moreover, how does a novice numerical taxonomist deduce which of these methods to use? Unfortunately, there is not a simple answer to the problem. However, most numerical taxonomy studies of bacteria involve use of the S_J and/or S_{SM} coefficients, which appear to generate acceptable results. We will not debate the often heated arguments as to which of these coefficients is 'best'. Nevertheless, the S_{SM} coefficient usually enables the delineation of more clearly defined clusters of OTUs.

Conversely with the S_J coefficient, clusters are more diffuse, with OTUs joining together at much lower S–levels. Yet, this coefficient is particularly useful in comparing unreactive organisms, such as *Flexibacter* spp., which might otherwise cluster together on the basis of characteristics which the OTUs did not possess, i.e., negative correlation. Consequently, the S_J coefficient is useful for confirming the validity of clusters defined initially by use of the S_{SM} coefficient. Routinely, use of these two coefficients together with vigour and pattern statistics should suffice. The latter sometimes highlight seemingly misplaced OTUs, by permitting their recovery in natural groupings.

Again the choice of clustering algorithm will influence the classification. The effect of different cluster techniques on a data matrix for *Bacillus* strains, which was derived using squared Euclidean distance, is shown in Fig. 2.9. There is a pronounced difference in the classifications. Complete linkage and Ward's method produced the most clearly defined clusters but this was accompanied by increased distortion of the taxonomic structure. OTUs were being assigned to clusters when they would be more correctly recovered as single member clusters. At the other extreme, single linkage and centroid sorting resulted in considerable chaining and few distinct clusters.

The accuracy with which a dendrogram represents a similarity matrix can be assessed using the determination of cophenetic correlation (Sokal and Rohlf, 1962). In this procedure, a similarity matrix is generated in which the entries are recorded from the dendrogram, the highest similarity levels linking two OTUs being the figures used (see Fig. 2.2.). This second matrix is then compared to the first using the Pearson product-moment correlation coefficient(r). A figure of $r = 1.0$ would be ideal,

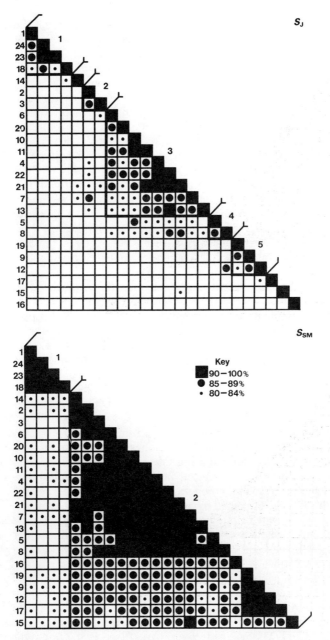

Fig. 2.8 The effect of different similarity coefficients on data for Enterobacteriaceae representatives. Clustering is by the unweighted average linkage algorithm.

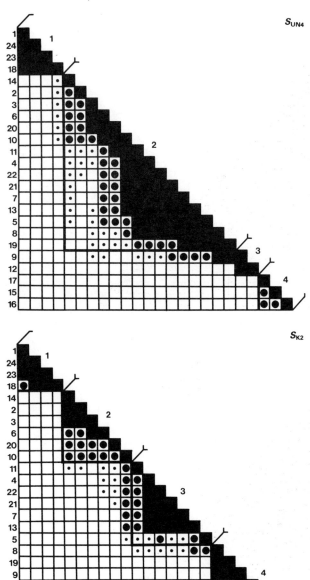

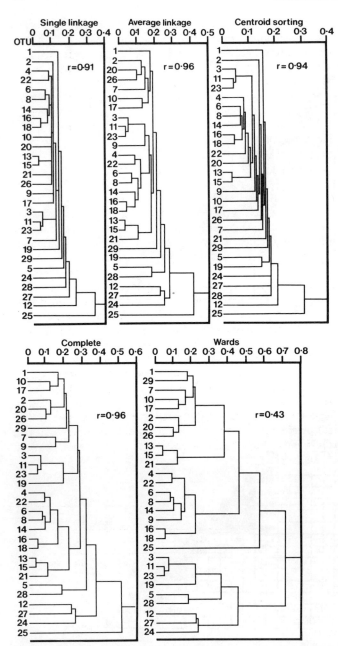

Fig. 2.9 The effect of clustering methods on a data matrix for *Bacillus* isolates, examined by squared Euclidean distance (*r* = cophenetic correlation; see text for details).

but since all 2-dimensional representations of taxonomic structure (which is multidimensional) must introduce distortion, this optimum cannot be obtained. Average-linkage cluster analysis generally gives the highest cophenetic correlation (Fig. 2.9; the example given in Fig. 2.2 is not representative because of the small sample of strains).

A second approach to the evaluation of a dendrogram is the determination of intra- and inter-group similarity values. These are the average similarities between OTUs of the same and of the different groups. The values provide an indication of homogeneity between OTUs of the same phenon, and of relatedness between all the phena.

Clusters of related OTUs (phena) may be further characterized by tabulating the percentage frequency of occurrence of each character among components of the phena (see Table 2.8). Thereafter these characters may be weighted (*a posteriori* weighting; i.e., weighting after the event) for inclusion in diagnostic schemes. These are necessary for the identification of further isolates. However, the identification of phena may be obtained from the presence of named reference cultures or by recourse to conventional schemes, such as typified by *Bergey's Manual of Systematic Bacteriology*. It may be expected that the outcome of any numerical taxonomy study will be the establishment of homogeneous taxonomic groups (taxa). Additional computational procedures may be used to determine representative OTUs from each phenon. Such 'average' organisms, which are referred to as hypothetical median organisms (HMOs), are suitable for allied studies, e.g., ecology.

Clearly, numerical taxonomy has gone a long way to improve the understanding of relationships between organisms, particularly facultatively anaerobic Gram-negative rods, based on overall similarity. However, there are some problems that need to be resolved, principally related to the lack of standardization of methodology and the reliance on conventional taxonomy for the identification of phena. Nevertheless,

Table 2.8. Characteristics of phenetic groups, in terms of percentage positive responses

Characteristics	Phena				
	1	2	3	4	5
Gram-positive	0	0	0	100	100
Presence of rods	100	100	100	100	0
Presence of cocci	0	0	0	0	100
Presence of endospores	0	0	0	100	0
Fermentatative metabolism	100	100	0	0	100
Oxidative metabolism	0	0	100	100	0
Catalase production	87	95	100	82	77
Oxidase production	71	83	75	23	0

these problems will probably be resolved, in which case the value of numerical taxonomy will be considerably enhanced.

REFERENCES

Alderson, G. (1985). The application and relevance of nonhierarchic methods in bacterial taxonomy, in *Computer-Assisted Bacterial Systematics* (M. Goodfellow, D. Jones and F. G. Priest, Eds.), pp. 227–263. Academic Press, London.

Austin, B. and Colwell, R. R. (1977). Evaluation of some coefficients for use in numerical taxonomy of microorganisms, *International Journal of Systematic Bacteriology*, 27, 204–210.

Clifford, H. T. and Stephenson, W. (1975). *An Introduction to numerical classification*. Academic Press, New York.

Dunn, G. and Everitt, B. S. (1982). *An Introduction to Mathematical Taxonomy*. Cambridge University Press, Cambridge.

Goodfellow, M., Austin, B. and Dickinson, C. H. (1976). Numerical taxonomy of some yellow-pigmented bacteria isolated from plants, *Journal of General Microbiology*, 97, 219–233.

Goodfellow, M., Embley, T. M. and Austin, B. (1985). Numerical taxonomy and emended description of *Renibacterium salmoninarum*, *Journal of General Microbiology*, 131, 2739–2752.

Logan, N. A. and Berkeley, R. C. W. (1981). Classification and identification of members of the genus *Bacillus* using API tests, in *The Aerobic Endospore-Forming Bacteria: Classification and Identification* (M. Goodfellow and R. C. W. Berkeley, Eds.), pp. 105–140. Academic Press, London.

Sneath, P. H. A. and Johnson, R. (1972). The influence on numerical taxonomic similarities of errors in microbiological tests, *Journal of General Microbiology*, 72, 377–392.

Sneath, P. H. A. and Sokal, R. R. (1973). *Numerical Taxonomy: The Principles and Practice of Numerical Classification*. W.H. Freeman, San Francisco.

Sokal, R. R. and Rohlf, F. J. (1962). The comparison of dendrograms by objective methods, *Taxon*, II, 33–40.

Wishart, D. (1978). *Clustan Users Manual*, Third Edition. Inter-University Research Councils Series Report No. 47, Program Library Unit, Edinburgh University.

3 Chemosystematics and molecular biology

3.1 INTRODUCTION

With the development of fast and reliable analytical techniques in chemistry and molecular biology, the reliance upon traditional micro-biological tests for gathering phenetic data has decreased. Instead, a wealth of new information has been obtained through chemosystematics or chemotaxonomy, defined as the study of chemical variations in living organisms, and the use of these chemical characters for classification and identification. This has broadened the database on which classifications are based. For some taxa, such as the actinomycetes, in which there are few useful physiological and morphological characters, chemosystematics, in particular, cell wall and lipid analyses, has been invaluable. For all bacteria, nucleic acid analyses are providing a sound taxonomic framework within which other approaches can be integrated. There have been suggestions in the past that chemotaxonomic data are in some sense more fundamental, or more closely a reflection of the genome than morphological traits, and will give more accurate classifications. Apart from nucleic acid sequences themselves, this does not seem to be the case, and biochemical, physiological and morphological classifications should, and do, resemble each other. In view of the connection between the two it would hardly be otherwise. Chemosystematics, therefore, provides us with a powerful supplement to the traditional approaches to classification and identification.

The levels within the cell at which chemotaxonomy operates are illustrated in Table 3.1, from which it is clear that chemical characteristics may be used to establish relationships at all levels within the taxonomic hierarchy. Thus DNA sequence comparisons and electrophoretic protein patterns are useful at the species level and, at the other extreme, ribosomal (r)RNA analyses are being increasingly used to delineate major divisions amongst all living organisms. This fine specificity, and yet ability to embrace wide taxonomic distance, is a particularly valuable aspect of these approaches.

Chemosystematic analyses are phenetic in nature. Relationships

Table 3.1 Chemosystematic analyses of the bacterial cell and the taxonomic level at which they are generally most useful

Cell component	Analysis	Taxonomic rank
Chromosomal DNA	Base composition (%GC)	Genus
	DNA:DNA reassociation	Species
Ribosomal RNA	Nucleotide sequence	
	Sequence catalogues	Genus and above
	DNA:rRNA hybridization	
Proteins	Amino acid sequence	Genus and above
	Serological comparisons	
	Electrophoretic patterns	Species and genus
	Enzyme patterns	
Cell walls	Peptidoglycan structure	
	Polysaccharides	Species and genus
	Teichoic acids	
Membranes	Fatty acids	
	Polar lipids	
	Mycolic acids	Species and genus
	Isoprenoid quinones	
Metabolic products	Fatty acids	Species and genus
Complete cell	Pyrolysis – gas liquid chromatography	Species and
	Pyrolysis – mass spectrometry	sub-species

between organisms are being assessed on the present-day structure of those organisms and the classifications obtained are phenetic, as, indeed are those derived from numerical analysis of the phenotype (see Chapter 2). In many instances, chemosystematic data may be processed using similarity coefficients and clustering or ordination algorithms to produce dendrograms and scatter diagrams analogous to those derived by traditional numerical taxonomy methods. However, analysis of the 'informational' macromolecules, DNA, rRNA and proteins may also be used to infer pathways of evolutionary descent if it is assumed that the diversity of types observed at present are derived from a single ancestor. This phylogenetic approach to classification using nucleic acid and protein sequences and the methods used to construct phylogenies will be discussed in Chapter 4, but the analyses themselves and the phenetic classifications so derived will be described below.

One of the major drawbacks of chemosystematics is the dependence of the chemical composition of microorganisms on the environment. Bacteria change their chemical composition substantially to accommodate environmental fluctuations. Thus, when comparing bacteria on the basis of some chemical component, it is important that the variation observed is a result of genetic differences and not due to an environmental effect. Cultures must, therefore, be grown under identical conditions and to the same stage of the batch culture growth cycle to ensure uniformity of environmental influence. This may be particularly difficult, sometimes impossible, if physiologically diverse organisms such as thermophiles and psychrophiles or aerobes and microaerophilic taxa are being compared.

Of the various chemical components used for taxonomy (Table 3.1), only chromosomal DNA and RNA are unaffected by growth conditions. The amounts of these molecules will fluctuate with growth rate, but the composition is invariant. Thus nucleic acids offer the only standard molecules by which the widest range of microorganisms (and higher eukaryotes) can be compared and classified. Constitutively synthesized proteins are also very useful in this respect but individual proteins may not be distributed universally. The cell walls of microorganisms vary widely in composition depending on the ionic complement of the medium. For example, phosphate limitation results in a total replacement of the phosphate-containing teichoic acids by teichuronic acid in the cell walls of many Gram-positive bacteria. The lipid composition of membranes is heavily influenced by temperature; at low temperatures unsaturated fatty acids tend to dominate and are replaced by saturated fatty acids at high temperature. And finally, the composition of the complete cell will vary according to the environmental conditions. It must, therefore, be remembered that, although chemosystematics has much to offer the bacterial taxonomist, in instances other than nucleic acid analyses, the cultural conditions must be carefully standardized and the possibility of environmental effects recognized.

3.2 CHROMOSOMAL DNA

The chromosome may be analysed at two levels. Firstly, the gross composition as the content of the four bases is of value. The duplex structure of DNA and hydrogen bonding between guanine (G) and cytosine (C), and between adenine (A) and thymine (T) bases, ensures the equivalent amounts of G + C and A + T are represented in the molecule. The content of the G + C (mol % G + C) in the molecule from bacteria varies within very broad limits from a minimum of about 25% (e.g., some mycoplasmas) to a maximum of approximately 75% (e.g., some streptomycetes. Beyond these limits the genetic code would be so skewed that few sensible proteins could be constructed.

Secondly although determination of the base sequence of DNA is now possible, it is not feasible for such large molecules as bacterial chromosomes. Nevertheless, an assessment of sequence homology between DNA molecules from two bacteria may be made by measuring the extent of renaturation of DNA molecules from the two sources in DNA 'reassociation' or 'pairing' experiments.

3.2.1   Methods for determining the base composition of DNA

Perhaps the most time-consuming aspect of all nucleic acid analyses is the preparation and purification of the material. Methods for the isolation of bacterial DNA are described in detail elsewhere (e.g., Owen and Pitcher, 1985). Essentially, cells are lysed with a detergent, such as sodium dodecyl sulphate. Gram-positive cells have first to be digested with lysozyme. Protein is removed by digestion with a non-specific protease (e.g., pronase) and chemical deproteinization using phenol or chloroform. RNA is removed with RNAase, and DNA can be selectively precipitated in the presence of ribooligonucleotides by isopropanol. DNA is concentrated and further purified by precipitation in ethanol, although carbohydrate contamination sometimes persists. This may be removed by hydroxylapatite chromatography or treatment with concanavalin A. It is important that the DNA is essentially free from protein, RNA and carbohydrate, because these impurities can interfere in subsequent assays for determining base composition or reassociation.

Direct estimation of the nucleotide composition of DNA involves hydrolysis followed by separation and quantitation of the products by chromatography. Although the introduction of high performance liquid chromatography (HPLC) has made such procedures more rapid and accurate, they remain less popular than the physicochemical approaches. These measure some physical parameter of the duplex molecule and relate this to mol % G + C using empirical formulae. A favoured approach is isopycnic equilibrium centrifugation in caesium chloride gradients. Centrifugation of a concentrated CsCl solution at high speed results in a density gradient. If a sample of DNA is incorporated into the CsCl solution, as the gradient is generated, so the DNA migrates to form a band at a point on the gradient that is equal to its own density (the isopycnic point). The buoyant density of the unknown can be calculated from the position of this band relative to a standard DNA of known buoyant density. There is a direct relationship between buoyant density and mol % (G + C), and thus the G + C content can be calculated [buoyant density = $1.66 + 0.098$ (G + C)]. This method is accurate and relatively insensitive to contamination of the DNA by protein or RNA, but the expense and complexity of the analytical ultracentrifuge have resulted in its declining popularity.

A second physical parameter of DNA related to base composition is the temperature at which the two strands separate. Thermal denaturation of duplex DNA is accompanied by an increase in absorbance at

260 nm (the absorbance maximum for DNA). A preparation of native DNA will increase by about 40% at A_{260} nm during denaturation—the hyperchromic shift (Fig. 3.1). The mid-point of the hyperchromic shift is referred to as the melting temperature (T_m) and is linearly related to mol % G + C. As with buoyant density, T_m can be converted to mol % G+ C using an established empirical formula; G + C (%) = 2.44 T_m − 169. Thermal denaturation is a popular method for the determination of base composition, since it is relatively insensitive to RNA or protein contamination of the DNA, and is rapid and inexpensive.

Both physicochemical procedures require comparison of an unknown against a standard or reference DNA. This is usually *Escherichia coli* DNA (51 mol % G + C) but, because different laboratories use different values for *E. coli* DNA, care must be taken in comparing results.

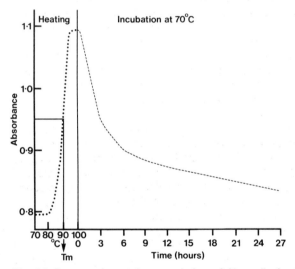

Fig. 3.1 Denaturation and reassociation of *Nocardia farcinica* DNA. Note the change in scale of abscissa from temperature to time (Adapted from Bradley and Mordarski, 1976).

3.2.2. Taxonomic value and applications of DNA base composition

Bacteria differ widely in base composition, but the G + C content is constant for a given organism. Similarity in base composition between two organisms does not necessarily imply relatedness, e.g., the G + C ratio of *Spirochaeta halophila* and *Pseudomonas testosteroni* is 62%. Since base composition does not take into account the linear sequence of bases in DNA, two organisms may have identical base compositions, but very few, if any, sequences and hence proteins in common. However, the converse is applicable. If two organisms possess DNA with widely different base composition, they will have few DNA sequences in common and are likely to be distantly related. Base composition is, therefore, a negative criterion; differences in base composition signify differences in nucleotide sequence in the DNA and hence dissimilar organisms.

It follows that DNA base composition provides a useful measure of genetic heterogeneity. If a genus contains species that differ widely in base composition, for example, the aerobic, endospore-forming bacteria of the genus *Bacillus* have DNA which varies from 32–67% G + C, it may be shown theoretically that organisms such as *B. cereus* (33% G + C) and *B. stearothermophilus* (53% G + C) can have few, if any, DNA sequences in common and should probably be allocated to different genera. Indeed from data gathered since 1960, it has been suggested that a useful guideline might be that the maximum genetic variation permissible within a genus should be that represented by 10–12% G + C. Thus genera such as *Bacillus, Lactobacillus* (32–52%), *Flavobacterium* (31–68%) and *Haemophilus* (37–55%) probably each require division into several genera. Most bacterial genera, however, have comparatively narrow ranges of G + C values which conform to this overall scheme (see Table 3.2).

Base composition of DNA is also useful at the species level. It has been suggested that members of a species should differ by no more than 5% G + C; deviation beyond this limit being indicative of excessive genetic dissimilarity. Differences in base composition within this 5% variation can be valuable, but it must be remembered that methods for determination of mol % G + C are seldom sufficiently accurate to distinguish reliably less than 2% variation. However, *Bacillus megaterium* strains have been shown to comprise two groups from estimates of base composition; 37–38% and 40–41%. Subsequent numerical taxonomy analyses revealed that two phenetically different species-ranked taxa were rep-

Table 3.2 Base compositions of some common bacterial genera

Genus (Gram-negative)	Mol % G + C	Genus (Gram-positive)	Mol % G + C
Acetobacter	51 – 65	*Bacillus*	32 – 67
Aeromonas	57 – 63	*Clostridium*	23 – 43
Cytophaga	29 – 45	*Corynebacterium*	48 – 59
Enterobacter	52 – 60	*Lactobacillus*	32 – 52
Escherichia	50 – 56	*Micrococcus*	68 – 73
Haemophilus	37 – 55	*Mycobacterium*	62 – 70
Klebsiella	52 – 56	*Staphylococcus*	30 – 37
Pseudomonas	57 – 70	*Streptococcus*	34 – 46
Rhodospirillum	60 – 67	*Streptomyces*	69 – 77

resented by these strains. Thus DNA base composition is a useful taxonomic tool to the extent that it is required information for the description of any new bacterial taxa.

It is tempting to speculate as to the reason for the heterogeneity of base composition amongst microorganisms which covers some 50% for bacteria, 30–70% for yeasts and fungi, 22–66% for protozoa, and 36–68% for algae. Given a typical bacterium with a $G + C$ content of 42% and average protein composition, it may be shown that the redundancy of the genetic code would allow limits of 31% and 67% $G + C$ to code for the proteins of the organism. However, there is slight variation in the protein composition of code-limit species, and, allowing for this, it can be calculated that their theoretical extremes would be 28% and 72%. These are very close to the observed extremes. Presumably there is some selection pressure to extremes of $G + C$, but the simple schemes of high ultraviolet environments selecting for high $G + C$ chromosomes (lack of $A + T$ for the induction of thymine dimers) or high temperature environments selecting for more thermostable (high $G + C$) chromosomes are not consistent with the observed patterns. Alternatively, it may be a random process. It is to be hoped that studies of gene expression in heterologous hosts will shed more light on codon usage in various bacteria.

3.2.3 Methods for calculating DNA sequence homology

Assessment of base sequence similarity between DNA from two organisms may be readily achieved by DNA reassociation experiments, in which DNA is rendered into single strands by thermal or alkali denaturation and subsequently allowed to anneal in the presence of a second denatured DNA molecule. If the nucleotide sequences of the two DNA samples are homologous, hybrid duplexes will be formed by base pairing. If there are few sequences in common, there will be negligible hybrid formation (Figs. 3.2 and 3.3). This technique, therefore, provides a quantitative estimate of DNA sequence homology between two organisms.

There are essentially two approaches to the determination of DNA base sequence complementarity; free solution assays (Fig. 3.2) and immobilized DNA assays (Fig. 3.3). The key feature is that a large excess (at least one thousandfold) of single stranded (ss)DNA is sheared to a constant molecular weight and incubated with radioactively labelled (usually ^{14}C) sheared ssDNA from a reference strain to allow renaturation to occur. Controls, comprising the homologous reaction and labelled DNA with salmon sperm DNA (which has no homology with bacterial DNA), are included. After renaturation, it is necessary to separate the double stranded (ds)DNA from remaining ssDNA. This may be achieved by treating the incubated mixture with a nuclease from *Aspergillus oryzae* (SI nuclease) that hydrolyses ssDNA but not dsDNA. After digestion,

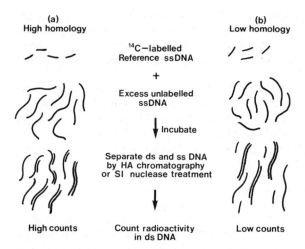

Fig. 3.2 Schematic representations of free solution DNA reassociation assays in which labelled reference DNA is incubated with an excess (at least 1000-fold) of unlabelled DNA from a second organism.

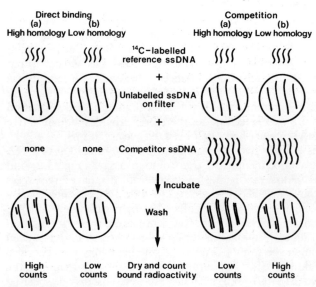

Fig. 3.3 Schematic representation of immobilized DNA reassociation assays performed either by direct binding or with competitor DNA. See text for details.

intact DNA is precipitated with trichloracetic acid, filtered and the radioactivity in the precipitate counted (Fig. 3.2). Percentage homology values are obtained by dividing the counts per minute in the heterologous SI-resistant DNA by the activity of the homologous reaction, and multiplying by 100.

An alternative scheme for separating the ds from ssDNA is hydroxylapatite chromatography. In phosphate buffers, only dsDNA will bind to hydroxylapatite, ssDNA is not adsorbed. The dsDNA can then be eluted with high molarity buffer, precipitated and counted in a scintillation counter. As before, percentage homology is the amount of label incorporated in the heterologous reaction divided by that in the homologous reaction, and multiplied by 100.

Free solution reassociation assays can be conducted without recourse to labelling DNA. The reassociation of ssDNA is accompanied by a reduction in A_{260} nm (Fig. 3.1). The initial rates of reassociation of an equimolar mixture of DNAs from two organisms are compared with the initial rates of the two homologous reactions. Non-homologous DNAs will reassociate more slowly (if at all) than the homologous molecules, and the relative rates may be used to determine percentage of homology from established formulae. This method has the advantage of being rapid; labelled DNA is not required and comparisons may be made between any pair of organisms rather than between reference and test strains. Moreover with due care and attention, it is as accurate as other procedures (see Owen and Pitcher, 1985).

Immobilized DNA reassociation assays use ssDNA bound to a solid support to effect the separation of renatured dsDNA (Fig. 3.3). Single stranded DNA will bind to nitrocellulose filters in such a way that the molecules retain the ability to form a duplex through hydrogen bonding of complementary sequences. In the direct assay, filters, containing immobilized ssDNA from different organisms, are incubated with labelled reference ssDNA to allow renaturation to occur. The membranes are subsequently washed, dried, and counted in a scintillation counter. The extent of homology is calculated as for the free solution assays described above, and high counts bound to the filter indicate high homology.

Competition reassociation is a variation of the direct immobilized assay and has the advantage that only reference DNA is bound to the membranes. This improves the accuracy of the method insofar as there is less variability in the amount of DNA bound to filters and it also reduces the laboratory workload. In competition experiments, a small amount of labelled reference ssDNA competes for reassociation sites on the homologous reference ssDNA immobilized on the filter with an excess of heterologous competitor ssDNA. If the competitor DNA is complementary to the reference DNA, it will preferentially reassociate with the bound DNA because of its higher concentration. Conversely, if there is little or no homology between the competitor and immobilized DNA molecules, the labelled reference DNA will reassociate with the membrane-bound DNA. Thus low counts on the washed dried membrane indicate high homology, and high counts represent low homology between the reference and heterologous competitor DNAs (Fig. 3.3). Methods for DNA reassociation have been fully described by Owen and Pitcher (1985).

All DNA reassociation assays must be carefully standardized if they are to give reproducible results since the extent and specificity of reassociation is heavily influenced by external conditions and the physical state of the DNA. There are essentially five factors that affect DNA reassociation assays (Brenner, 1970):

1. DNA reassociation is influenced by the size of the DNA fragments; the larger the fragment then the greater the rate of reassociation in free solution, but, as the viscosity of the solution increases, the rate of the reaction decreases. Moreover, very small fragments (less than about 15 nucleotides in length) show little specificity during renaturation; a compromise is therefore required. It it usual to reduce the size of the fragments by physical shear using sonication or pressure drop in a French press to a uniform molecular weight of $1 - 5 \times 10^5$ daltons, representing about $200 - 1000$ base pairs.

2. The rate and extent of renaturation increases as the ionic strength of the incubation buffer is increased. Duplexes formed at high ionic strength are less temperature stable than those formed at lower ionic strength, indicating that a loss of specificity in pairing occurs at the higher ionic strength. It is, therefore, important to use standard buffers for these experiments.

3. Purity of the DNA preparation is important; contamination with RNA or carbohydrates can interfere in the extent of reassociation. It is also important to ensure that the DNA samples are completely denatured into single strands.

4. DNA concentration and time of incubation are critical features in reassociation assays. Britten and Kohne (1966) introduced the concept of 'cot' and showed that duplex formation is a function of the initial DNA concentration (C_0) and the time of incubation (t). Cot can be readily calculated as $\frac{1}{2}$ $(A_{260}$ nm) multiplied by the incubation time in hours. An example of a Cot curve is shown in Fig. 3.4; the sigmoid form of the graph indicates that reassociation is a second-order reaction, and it is evident that reassociation is complete by 100 Cots.

   In free solution assays, it is impossible to distinguish between renatured labelled DNA and the hybrid comprising labelled and non-labelled DNA. By using a 4000-fold excess of unlabelled

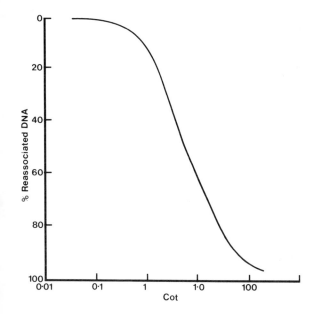

Fig. 3.4 Time course of reassociation of *E. coli* DNA. Labelled and unlabelled DNA fragments from *E. coli* were incubated to various Cot values and the amount of reassociation determined (see text for details; adapted from Brenner, 1970).

DNA (i.e., 0.1 µg ml^{-1} labelled DNA and 400 µg ml^{-1} unlabelled DNA) incubated for 21 h, the labelled DNA receives a Cot of about 100 and the labelled DNA a Cot of 0.025, which would result in about 1% reassociation of labelled DNA. For immobilized DNA assays, a lower ratio (100:1) is usually sufficient to ensure sufficient sites on the filter for the labelled DNA to reassociate, and the homologous labelled DNA is not a serious problem.

5. The temperature, at which the reassociation mixture is incubated, is critically important. Optimal renaturation occurs at about 30°C below the T_m of the DNA mixture. Below this temperature, non-specific hybrids are formed and distantly related sequences reassociate. Above this temperature, the pairing is more stringent, to the extent that the reaction is reduced considerably and only closely related sequences form stable duplexes. It is usual to work at 30°C below T_m for general studies, but, if closely related bacteria are being examined, the more stringent 15°C below T_m is used.

By careful attention to these parameters, DNA reassociation studies are accurate and repeatable and, despite the variety of procedures available, congruent results are obtained.

3.2.4 Taxonomic value and applications of DNA reassociation

DNA reassociation is recommended for the evaluation of phenetic relationships for several reasons:

1. The DNA composition of a cell is invariant regardless of the growth conditions, although the content will vary with growth rate. Thus the stage of growth (lag, exponential, or stationary phase, even sporulated cells) or nature of the medium will have no effect on the reproducibility of the results. As a consequence, classifications based on DNA reassociation tend to be more stable than those derived from other procedures.

2. A second feature of DNA reassociation, that contributes to the stability of the classifications it provides, is that the estimation of relatedness between organisms is based on the complete genotype. Numerical phenetics compares organisms on the basis of a fraction of the genotype (perhaps 5 - 20%) and this is valid so long as a representative sampling of the genome is achieved. This is attempted by analysing a heterogeneous battery of characters and, in general, the aim of representative sampling is presumably achieved since classifications from DNA reassociation and numerical phenetics are largely congruous. Data for % similarity and DNA reassociation for some enterobacteria is shown in Fig. 3.5. The relationship between numerical phenetics and % DNA pairing is not a straight line since there is very little sequence homology at phenetic similarities below about 50% S_{SM}. The presence of apparent phenetic similarity at low DNA sequence homology probably results from negative matches [absence of a gene(s)] which are included by the S_{SM} coefficient (see Chapter 2) as a measure of similarity and yet unlikely to be recognized as shared DNA sequences. The S_{SM} coefficient may therefore give rise to erroneously high measures of similarity. Even with the S_J coefficient, which ignores negative correlations, the shared presence of an attribute (e.g., gas production from glucose, or resistance to an antibiotic) may be due to entirely different enzymes and consequently different DNA sequences. Moreover, the same protein may be encoded by completely different DNA sequences because of the redundancy of the genetic code. The reverse situation, in which homology values are higher than phenetic similarity values, is not so pronounced. It probably results amongst highly related bacteria from slightly mismatched

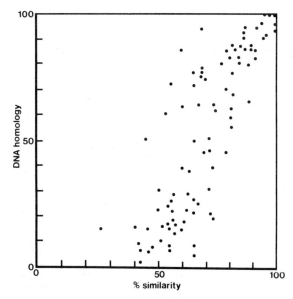

Fig. 3.5 Correlation of % DNA sequence homology and % similarity (phenetic) (Adapted from Staley and Colwell, 1973).

duplexes formed under optimal conditions being estimated as 100% reassociated. Areas of localized mismatched bases, or deletion mutations, are not detected by the relatively crude methods used for processing large numbers of reassociation determinations, and yet could be responsible for lack of a particular attribute and estimated as a difference in phenotype by numerical taxonomy. It is, however, encouraging that there is this general congruence between DNA reassociation and numerical phenetics that lends credibility to both methods.

3. DNA homology is providing a unifying concept of the bacterial species. A bacterial species is usually defined as a group of strains that share many features, and differs considerably from other groups. This definition is subjective, and, it is not surprising that some species are far more heterogeneous than others. A more objective and rigorous definition of species is desirable, and DNA homology studies are indicating how this might be achieved. By examining DNA reassociation data gathered since the techniques were introduced in the early 1970s, it has become apparent that organisms within well defined species have extensive DNA sequence homology. Moreover, organisms representing different species have very few sequences in common.

The lower allowable limit of sequence homology within a species is a controversial topic, but between 50 and 60% reassociation under optimal conditions seems a reasonable compromise. This signifies the great practical value of the technique of which just one example must suffice.

Bacillus circulans has long been a problematical species encompassing a diverse range of phenotypes. Indeed, it has usually been described as a 'spectrum' since many different phenotypes existed, but no clear distinctions were apparent. DNA reassociation studies have recently revealed at least five DNA homology groups, which have been substantiated by phenotypic tests, and, thus, new species have been described.

4. DNA reassociation analyses may also be used to deduce evolutionary pathways amongst closely related organisms. This is achieved by estimating DNA sequence divergence from a supposed ancestral type using thermal stabilities of reassociated duplexes. Hybrid DNA molecules generally have a lower T_m than parental homologous molecules resulting from mismatched sequences; each $0.7°C$ drop in T_m represents 1% mismatched bases. By measuring the reduction in T_m (ΔT_m) for a hybrid duplex, it is possible to estimate the extent of sequence divergence. Thus, a hybrid duplex formed at $30°C$ below T_m between *E. coli* and *Serratia marcescens*, would reassociate about 25%. The ΔT_m would be $13-14°C$, indicating about 20% sequence divergence in the reassociated sequences. Conversely, DNA from *E. coli* and *Shigella dysenteriae* would reassociate 80–90% with a ΔT_m of $1-6°C$ representing only some 4% mismatched sequences. This provides an estimate of affinity between very closely related bacteria and, if *E. coli* were the common ancestor, it can be argued that *Ser. marcescens* diverged earlier in evolutionary time than *Sh. dysenteriae*.

An alternative to ΔT_m is to measure % reassociation at optimal and stringent temperatures, and divide the latter figure by the former. This ratio, the divergence index or thermal binding index, approaches 1.0 in closely related bacteria that have few sequence dissimilarities, and 0.00 in unrelated bacteria. It is less laborious than ΔT_m determinations.

DNA reassociation techniques also have their limitations. Firstly, they do not lend themselves to rapid, automated identification procedures in the same way as phenotypic tests. Hybridization probes for rapid detection and identification of viruses and specific genes within bacteria are being developed (see Chapter 6) but, at present, are not readily applicable to most bacteria. Secondly, because of the amount of work involved, full similarity (reassociation) matrices with estimates of

DNA homology between each and every strain are seldom produced. Instead, comparisons are made between judiciously chosen reference strains and a variety of test strains. Generally, this approach is reliable but the reference strains must represent the taxa under study. If they do not, distortion of the taxonomic structure may occur. It is therefore, prudent to use DNA reassociation in conjunction with a second approach, such as traditional numerical taxonomy. The latter would provide boundaries for taxa, which can then be validated using DNA reassociation based on sequence homology with centrotype strains as references. Numerical taxonomy would also provide the diagnostic features for the construction of identification schemes. This combination is a powerful and reliable approach to bacterial classification and identification.

3.3 ANALYSIS OF RNA

There are three categories of RNA in prokaryotes; the short-lived messenger (m)RNA responsible for transmitting information from the chromosome to the ribosome, and the stable forms, i.e., transfer (t)RNA, which decodes the message, and ribosomal (r)RNA involved in the structure of the ribosome and the reading of the message. Analysis of RNA for taxonomic purposes focuses on the three rRNAs; the 5S, 16S and 23S molecule. The value of these molecules as indicators of relatedness is severalfold:

1. The rRNAs are essential elements in protein synthesis and are, therefore, present in all living organisms (with the notable exception of viruses).
2. Because of the conserved functions of these molecules, they have changed very little during evolution. Thus rRNAs from even the most taxonomically distant organisms, that share virtually no DNA sequence homology, will have rRNA sequences in common, and, therefore, relatedness can be assessed. Ribosomal RNA is probably unique amongst macromolecules in this respect.
3. Some segments of rRNA have evolved more rapidly than others allowing comparisons amongst relatively closely related organisms.
4. If it is assumed that bacteria have evolved from a common ancestry, phylogenetic lines of descent may be inferred from rRNA sequences (see Chapter 4).

Methods for determining similarity between rRNA molecules are of three kinds. Since the 16S and 23S molecules are too large for primary sequence determination on a routine scale (about 1600 and 3300 nucleotides, respectively), this approach is largely restricted to the 5S molecule. The larger molecules can be compared by comparative cataloguing however, in which the rRNA is digested with ribonuclease T, and the sequences of the resultant oligonucleotides are compared. This method has been used extensively for the 16S molecule, and the third approach, DNA:rRNA hybridization, has been used for both 16S and 23S molecules with similar results.

3.3.1 Sequence analysis of 5S rRNA

Prokaryotic 5S rRNAs are comprised of 120 nucleotides (from Gram-negative bacteria) or 116 (sometimes 117) nucleotides as found in Gram-positive organisms. Sequences may now be determined routinely using methods similar to the Maxam – Gilbert procedure developed for DNA sequencing. After radioactively labelling an end of the molecule, it is partially digested with base specific enzymes or reagents and the products are separated in four 'base specific' lanes by polyacrylamide gel electrophoresis (PAGE). This generates a ladder from which the base sequence can be read. To date, some 175 5S rRNA sequences have been determined (reviewed by Pace *et al.*, 1985).

It is necessary to align the sequences for comparisons to be made. The 120 and 116 nucleotide molecules can be aligned by inserting two residue gaps into the 5' and 3' ends of the 116 nucleotide molecule. Alignment with eukaryotic (120 nucleotide) molecules is more complex, and involves estimating the 'best match alignment' with minimal gap insertions. Simple comparisons of the molecules in terms of sequence homology can be clustered to provide a dendrogram (Fig. 3.6). Alternatively, more

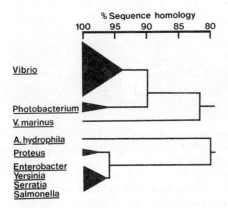

Fig. 3.6 Dendrogram indicating relationships amongst genera of the Enterobacteriaceae and Vibrionaceae based on 5S rRNA sequence data (from MacDonell and Colwell, 1984, with permission).

complicated comparisons can be made that attempt to account for mutation and back mutation rates (see Chapter 4).

3.3.2 Comparative cataloguing of rRNA

It can be argued that the small size of the 5S rRNA molecule detracts from its value in measuring relatedness between organisms, since it can undergo marked mutational change that would be obscured by the long stretches of conserved sequence present in the larger molecules. Largely for this reason, Woese and his collaborators chose the 16S molecule, which is somewhat easier to handle than the 23S for the development of comparative cataloguing (Stackebrandt and Woese, 1981). This process involves digesting purified 16S rRNA with T1 ribonuclease into oligonucleotides. The 5′-terminus of these molecules is labelled *in vitro* with ^{32}P, and they are separated by two-dimensional paper chromatography or, more recently, thin layer chromatography (TLC). This provides an oligonucleotide 'fingerprint' of the rRNA and each oligonucleotide is then sequenced to produce a 'catalogue' of sequences. Taxonomic structure is derived from catalogues of organisms, by comparing each catalogue with every other catalogue. Oligonucleotides of six residues or more, common to two catalogues, are considered and an estimate of similarity is calculated using a Dice-type coefficient defined as:

$$S_{AB} = 2N_{AB}/(N_A + N_B)$$

where N_{AB} is the total number of residues in common amongst the oligonucleotides from the two organisms A and B, N_A is the total number of residues in oligonucleotides from organism A, and N_B is the total number of residues in the oligonucleotides from organism B. Thus S_{AB} values, like other coefficients of affinity, range from 1.0 for identical molecules to about 0.03 for totally unrelated molecules. The S_{AB} values can be arranged as a similarity matrix and analysed by standard clustering procedures (see Chapter 2), usually average linkage analysis, to produce a dendrogram (Fig. 3.7). We emphasize that this dendrogram shows the present day relationships of organisms based on the 16S rRNA sequence and is a phenetic classification, just as a numerical taxonomy analysis of the phenotype is phenetic. Phylogenetic inferences can be extracted from this dendrogram if so desired (see Chapter 4).

With current developments in RNA sequencing, it is now possible to determine the nucleotide sequence of large stretches of RNA, and it may well be that cataloguing will be superceded by complete sequence analysis but, to date, only about 20 complete 16S RNA sequences are known (Pace *et al.*, 1985).

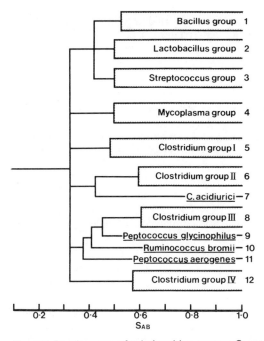

Fig. 3.7 Dendrogram of relationships among Gram-positive bacteria based on comparative cataloguing of 16S rRNA (adapted from Stackebrandt and Woese, 1981). *Bacillus* group: *Bacillus* species, *Peptococcus saccharolyticus, Planococcus citreus, Sporolactobacillus inulinus, Sporosarcina ureae Staphylococcus* species, *Thermoactinomyces vulgaris. Lactobacillus* group: *Kurthia zopfii, Lactobacillus* species, *Leuconostoc mesenteroides, Pediococcus pentosaceus. Streptococcus* group: *Streptococcus* species. *Mycoplasma* group: *Acholeplasma laidlawii, Clostridium innocuum, C. ramosum, Mycoplasma capricolum, M. galli, Spiroplasma citri. Clostridium* group II: *C. litus-eburense, Eubacterium tenue, Peptostreptococcus anaerobius. Clostridium* group III: *C. sphenoides, C. aminovalericum. Clostridium* group IV: *C. barkeri, Acetobacterium woodii, Eubacterium linosus.*

3.3.3 DNA:rRNA hybridization

Ribosomal RNA is transcribed from about 10 cistrons in the chromosome. In DNA:rRNA hybridizations, the sequence homology between labelled 16S or 23S rRNA from a reference strain, and the rRNA cistrons within the chromosomal DNA from a second organism, are determined. This estimate of homology can be further analysed by estimating the extent of mismatched bases from the depression in T_m of the hybrid (see Section 3.2.4).

The usual approach is to immobilize denatured chromosomal DNA (about 50 μg) from a range of bacteria on nitrocellulose filters. The filters are incubated with ^{14}C-labelled rRNA (10 μg) at optimal hybridization temperature for 16 h. The filters are then washed, treated with RNAase, and the total amount bound (as μg RNA per 100 μg DNA), is the 'percent rRNA binding'. The thermal stability of the hybrid is estimated by measuring the release of label as the temperature of the filter is increased, and the $T_{m(e)}$ is the temperature at which 50% of the hybrid is eluted.

Taxonomic relatedness is revealed in two ways. Similarity maps, in which the ordinates are % rRNA binding and $T_{m(e)}$, reveal clusters of organisms and show overlap between groups but do not provide a hierarchical classification. Where many reciprocal $T_{m(e)}$ determinations have been performed, the data can be treated as in a similarity matrix and clustered using the usual algorithms to provide a dendrogram that indicates the relationships in a hierarchical fashion (Fig. 3.8).

3.3.4 Value and applications of rRNA studies

Ribosomal RNA analysis is an invaluable procedure for extending classifications, established by DNA reassociation and phenotypic studies, to more distant organisms, and is being used to provide a comprehensive view of the relationships among all prokaryotes. Of the three methods, sequencing 5S RNA and cataloguing 16S RNA molecules have the advantage of providing information for individual organisms that can be

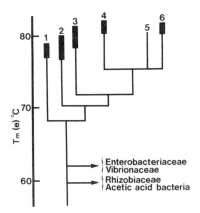

Fig. 3.8 Dendrogram of relationships among some Gram negative bacteria based on thermal stabilities of rRNA: DNA hybrids. The shaded areas represent the range of $T_{m(e)}$ values of the reference taxon.
1, *Janthinobacterium lividum*; 2, *Chromobacterium violaceum*; 3, *Pseudomonas solanacearum*; 5, *Alcaligenes denitrificans*; 6, *Alcaligenes faecalis* (adapted from De Ley *et al.*, 1978).

processed using estimates of similarity and clustering algorithms. Conversely, these approaches are expensive in time and materials, which severely limits the number of organisms that can be examined. Improvements in sequencing technology, however, are sufficiently rapid that these methods will become the most popular. Indeed, natural microbial populations can now be analysed *in situ* by providing a surface within an ecological niche for the growth of a colony, removing the biomass and directly isolating and sequencing the 5S RNA to provide identifications (Pace *et al.*, 1985).

Hybridization studies suffer the disadvantage that they do not generate data for an individual strain, but compare sequences between reference and test strains. Moreover, hybridization is not so accurate as sequencing and cataloguing, particularly amongst distantly related organisms. Nevertheless, it is rapid, relatively straightforward and has provided valuable insight into the relationships amongst several Gram-negative and Gram-positive taxa (Gillis and De Ley, 1980).

The application of rRNA analysis is having a dramatic effect on our more traditional views of the relationships amongst the prokaryotes. Without resource to phylogenetic considerations, phenograms based on rRNA sequences (either determined from 5S rRNA sequences, 16S rRNA catalogues or rRNA:DNA hybrids) are providing:

1. A comprehensive overview of the classification of prokaryotes and indicating relationships that, on phenotypic grounds, had not been apparent, and
2. Unifying concepts of the genus and higher-ranked taxa.

With regard to the first point mentioned above, some 400 bacterial species have now been characterized by 16S rRNA cataloguing (Stackebrandt and Woese, 1984) and one of the most fascinating implications of this study is that prokaryotes fall into one of two fundamental groups, namely, the archaebacteria and the eubacteria. Bacteria from these groups have very little sequence homology in their rRNAs, which show S_{AB} values of 0.1 or lower, and also differ in several other important respects. Archaebacteria share some interesting features with eukaryotes; their tRNA genes have introns and their protein elongation factors are sensitive to diphtheria toxin. Unlike eubacteria, archaebacteria do not contain peptidoglycan, and their membrane lipids do not comprise the usual ester linked fatty acids but instead are glycerol ethers involving phytane side-groups. Moreover, archaebacterial RNA polymerases are of a different structure from the eubacterial enzymes. Currently, the archaebacteria are considered a major division within the kingdom Procaryotae and include halophiles, methanogens and the sulphur-dependent organisms, formerly termed thermoacidophiles.

In addition to revealing this fundamental division of the procaryotes, rRNA cataloguing has also uncovered some initially surprising bacterial relationships. For example, endospore-formation may not be as important as once thought, since the *Bacillus* group, based on rRNA cataloguing, contains the asporogenous genera *Staphylococcus* and *Planococcus*, but not the anaerobic sporeformers of the genus *Clostridium* (Fig. 3.7). Amongst Gram-negative bacteria, photosynthesis seems to have been a particularly misleading character, since *Rhodopseudomonas* and *Rhodospirillum* are members of separate taxa, both of which contain non-photosynthetic types. Similarly, undue importance on morphology can distort classification, since spiral bacteria such as *Spirillum*, *Aquaspirillum* and *Oceanospirillum* show little rRNA relatedness. Moreover, the endospore-forming filamentous bacteria of the genus *Thermoactinomyces* have been assigned to the *Bacillus* group on rRNA cataloguing and other criteria, rather than to the actinomycetes.

With regard to the ranking of taxa, rRNA studies, like DNA base composition analyses, have revealed immense discrepancies in the genetic heterogeneity contained within different genera. For example, *Clostridium* represents four groups each of which is more diverse than the complete *Aeromonas*, *Vibrio*, Enterobacteriacease group! *Bacillus* is also probably underclassified, where the streptomycetes are overclassified and many synonymous species have been described. Some genera, such as *Chromobacterium* and *Janthinobacterium*, have been proposed largely on the basis of rRNA:DNA hybridizations that revealed the individual taxa, and phenotypic studies subsequently provided distinguishing features. It seems likely that phenotypic information alone will be insufficient for any major revisions of bacterial taxa at the genus level or above, and rRNA analyses will become increasingly important. The phylogenetic implications of rRNA analyses are discussed in Chapter 4.

3.4 ANALYSIS OF PROTEINS

Measurement of relationships between organisms using proteins, either focuses on comparisons of individual molecules using amino acid sequences or serological cross-reactions, or gross evaluation of total cellular protein. A refinement of the latter approach is to compare the activities of sets of enzymes. These approaches and some example applications will be described below, with the exception of protein sequence analysis which is covered in Chapter 4.

3.4.1 Comparative serology of proteins

The development of quantitative serological techniques has enabled the rapid detection of structural similarities in isologous enzymes from different bacteria. A common approach is to use the chosen purified enzyme from several reference strains to raise antisera. These antisera are used to detect the isologous enzyme in cell extracts from test organisms. Two-dimensional immunodiffusion (Ouchterlony technique) is used to establish gross similarities between enzymes, a line of identity indicating closely-related molecules, and crossed precipitin lines revealing more distantly related enzymes. The degree of relatedness can be estimated using microcomplement fixation tests in which the amount of complement fixed by homologous and heterologous mixtures can be used to quantify the reaction. The figures obtained can be analysed using the usual clustering techniques.

It is generally assumed that sequence variation in proteins will be reflected in their secondary structures and, therefore, antigenicity. Quantitative serology is therefore a rapid approach to the analysis of protein structure in much the same way as rRNA:DNA hybridization is a rapid approach to rRNA sequencing. However, it has not been a particularly popular technique. Most studies have centred on the lactic acid bacteria; *Lactobacillus*, *Pediococcus*, *Leuconostoc* and *Streptococcus* and the enzymes fructose (bisphosphate) aldolase, and glucose-6-phosphate-, glyceraldehyde-3-phosphate- and lactate dehydrogenases. The results confirm the close relationship of these genera, and are largely in accord with classifications derived from other sources. It would seem that a particularly useful aspect of comparative serology is to provide the preliminary ground work, and to indicate taxa and proteins that can be analysed in detail using protein sequencing. Nevertheless, like rRNA:DNA reassociation, it is also a valuable approach to detecting molecular relationships in its own right.

3.4.2 Comparative electrophoresis of cellular proteins

The bacterial genome is largely devoted to the production of some 200 proteins that function either enzymically or structurally. When a bacterium is grown under carefully standardized conditions, this protein complement is essentially invariant. Electrophoresis of the total cellular proteins in polyacrylamide gels (PAGE) provides a partial separation in which individual bands mostly represent several proteins. However, this complex pattern is reproducible and represents a 'finger print' of the strain that can be used for comparative purposes (Fig. 3.9).

The original electrophoretic systems involved polyacrylamide rod gels used in non-denaturing conditions, but more recently sodium dodecyl sulphate (SDS) PAGE has found greater application. The bacteria are broken, usually by physical means, and the lysate applied directly to the gel. One or two reference proteins are included and, after electrophoresis, the gel is stained with, for example, Coomassie blue. A densitometer trace of the stained gel provides the quantitative data for the bacterium.

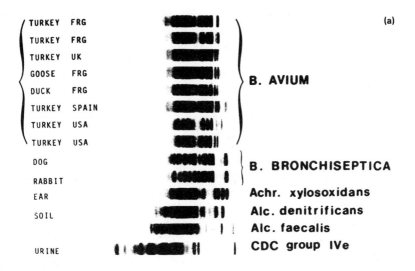

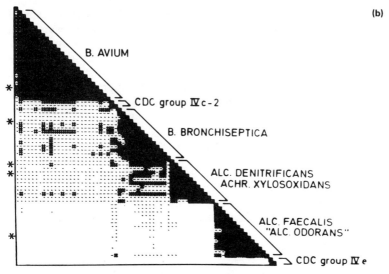

Fig. 3.9 (a) Normalized electrophoregrams of soluble proteins from 8 avian *Bordetella*-like strains (*B. avium*) and representative strains of various more or less allied bacteria. The origin of the strains is shown at the left side of the figure. The avian strains were isolated in the Federal Republic of Germany (FRG), United Kingdom (UK), Spain, and United States of America (USA). (b) Sorted similarity matrix from the numerical analysis of protein patterns of 24 avian *Bordetella*-like strains (*B. avian*) and various related strains: ■, 95–100% similarity; ◘, 90–94%; #, 85–89%; —, 80–84%; ●, 70–79%; □, 0–69% (From Kersters, 1985, with permission).

Analysis of the densitomer traces uses traditional numerical methods. The peak heights (absorbances) are normalized using the internal reference proteins after removal of background 'noise', and used as the 'characters' for the organism. Similarities are calculated between each organism using a suitable similarity coefficient; the Pearson product-moment correlation and Dice coefficients, have been popular. The resultant matrix is clustered using the average linkage algorithm to provide a sorted similarity matrix or dendrogram. Computer programmes are readily available to perform these calculations (Jackman, 1985).

Comparative electrophoresis of proteins should be equivalent to DNA:DNA reassociation, since bacteria are being compared on the translational products of most of the cell's chromosome. It is to be expected, therefore, that it will be most useful at the species level, and of little value in comparing distantly related bacteria. This is the experience to date. The two methods give highly congruent results. Protein electrophoresis has several advantages over DNA:DNA pairing. Clusters are formed from complete similarity matrices, since a set of data is produced for each strain rather than comparisons between reference and test strains. The technique is more rapid than DNA reassociation particularly since, 20 samples can be electrophoresed in a single slab gel, there are no lengthy preparation procedures and the data can be entered directly from the densitometer into a computer. Thirdly, it is more amenable to identification than DNA:DNA reassociation. Patterns for reference clusters can be held on disk, and patterns for unknown organisms can be compared with the data bank in order to provide an identification in much the same way as probabilistic methods are used for identification based on phenotypic characters (see Chapter 6). However there are some drawbacks. Although reproducibility within the laboratory does not seem to be a serious problem, inter-laboratory comparisons may not always be entirely reliable. The adoption of standard procedures may improve reproducibility. Secondly, although any group of bacteria can be studied, standardization of growth conditions may be difficult or impossible when comparing different physiological types.

Despite the advantages of the system, comparative electrophoresis of proteins has not been widely adopted. It has been used in conjunction with traditional numerical taxonomy and DNA:DNA reassociation to delineate species of *Zymomonas* and for the classification and identification of *Alcaligenes, Achromobacter, Bordetella, Corynebacterium* and *Pseudomonas* species. In all cases, there was excellent agreement between the three approaches (for review see Kersters, 1985).

3.4.3 Enzyme patterns

Rather than an analysis of the total proteins in a cell, the properties of one or more enzymes can be compared. Such studies have focused on enzymes that are common to a group and can be readily detected in electrophoresis gels through a specific reaction that results in colour formation. Esterases and dehydrogenases have been popular. Thus, strains can be classified into groups on the basis of the presence or absence of particular enzymes and their electrophoretic mobilities.

The value of this method is that it is rapid, indeed, it is often possible to produce a result within a few hours from an isolated colony. On the other hand, it is of doubtful value for creating a classification, since it is likely that minor modifications in these key enzymes could alter significantly their electrophoretic mobility. This approach is, therefore, a useful supplement to procedures based on numerical analysis of protein profiles, or of phenotype, or DNA:DNA reassociation and, in general, strains belonging to the same species possess similar zymograms (Williams and Shah, 1980).

Electrophoretic analysis of malate dehydrogenase from *Bacteroides* species has proved useful for the rapid identification of subspecies of *B. melaninogenicus* associated with periodontal disease, and lactate dehydrogenase patterns have been used for the classification of *Lactobacillus* species but, in general, this approach has not been used extensively and is being supplanted by alternative procedures, such as comparative electrophoresis of total proteins.

3.5 CELL ENVELOPE ANALYSES

Correlating with the different reactions to the crystal violet/iodine complex of Gram's stain, Gram-positive and Gram-negative organisms have very different cell envelopes. The Gram-positive bacterium is surrounded by a cytoplasmic membrane bounded by a thick layer of peptidoglycan, containing covalently bound teichoic acid. The myco-bacteria and other acid fast bacteria have a modified Gram-positive envelope in which considerable lipid material is intercolated into the peptidoglycan and they, therefore, stain Gram-negative. True Gram-negative bacteria possess a double-layered envelope comprising the cytoplasmic membrane, a thin layer of peptidoglycan within a hydro-philic compartment, termed the periplasm, and an outer membrane comprised of lipoprotein and lipopolysaccharide (Rogers, 1981). Structural variation in these various components and molecules can be used to classify and identify bacteria.

3.5.1 Peptidoglycan

Peptidoglycan (murein) is found in all bacteria, except mycoplasmas and archaebacteria. It comprises amino-sugar backbones bearing tetrapeptide chains with a diamino acid in position three. These chains are cross linked either directly or through an interpeptide bridge of 1–6 amino acid residues between the diamino acid and the D-Ala residue at position four on an adjacent chain. The animo sugar backbones are largely homogeneous, but the diamino acid of the tetrapeptide chain and the composition of the interpeptide bridge show variation between taxa. This information is especially useful for Gram-positive bacteria; Gram-negative bacteria are, however, remarkably uniform in peptidoglycan structure.

Data from qualitative cell wall sugar and amino acid analyses have been particularly valuable in the systematics of actinomycetes and related bacteria, which have been allocated to eight wall chemotypes. For example, the nature of the diamino acid proved particularly important for the classification of coryneform bacteria. *Corynebacterium diphtheriae* and related pathogenic strains have directly cross-linked *meso*-diaminopimelic acid-containing peptidoglycans (wall chemotype IV) like several pathogenic mycobacteria and nocardia, whereas other coryneform genera, such as *Arthrobacter* and *Microbacterium*, have the *meso*-DAP replaced by lysine and belong to wall chemotype VI (Goodfellow and Cross, 1984).

Methods for determining wall chemotype are fairly simple and rapid, being based on qualitative chromatographic analysis of acid hydrolysates of whole organisms or purified cell walls. More precise taxonomic data can be obtained from the primary structure of the peptidoglycan, but the analytical methods involved are specialized and beyond the scope of the average microbiological laboratory.

Polysaccharides and teichoic acids have received scant attention as chemotaxonomic markers.

3.5.2 Lipids

Lipids provide a wealth of taxonomic information that may be used for both classification and identification. There are three important classes; long-chain fatty acids, polar lipids and isoprenoid quinones.

Long-chain fatty acids can be released from polar lipids of the plasma membrane by esterification, and analysed by gas chromatography (GC). Variation in fatty acid composition includes the length of the carbon chain, which can range from about 8 to 26 carbons.

The length of the alkyl chain influences menbrane fluidity. Growth at higher temperatures promotes the synthesis of longer chains. Introduction of a double bond in the chain also affects membrane fluidity, but the degree of unsaturation is currently of doubtful usefulness in taxonomy. Hydroxylated fatty acids have the OH group in either position 2 or 3, and are found in almost all Gram-negative species (Table 3.3). Cyclopropane fatty acids are found in major quantities in both Gram-negative and Gram-positive bacteria, whereas fatty acids, with branched alkyl chains, predominate in Gram-positive genera. Finally, the high molecular weight 3-OH-α-branched fatty acids (mycolic acids) are found only in the complex lipids of *Mycobacterium*, *Corynebacterium* and *Nocardia* strains (Dobson *et al.*, 1985).

Table 3.3 Some Gram-negative genera grouped accordng to their content of hydroxy-, methyl- and cyclopropane substituted fatty acids (after Jantzen and Bryn, 1985).

FA group	OH-FA	BR-FA	CYCLO-FA
Acinetobacter, Bordetella pertussis, Haemophilus, Moraxella, Neisseria, Pasteurella	+	–	–
Alcaligenes, Enterobacteriaceae. *Flavobacterium-Cytophaga* 1.	+	–	+
Bacteroides, Flavobacterium-Cytophaga 2.	+	+	–
Legionella.	–	+	–
Brucella.	–	–	+
Treponema, Gardnerella.	–	–	–

Abbreviations: FA, fatty acid; OH-, hydroxylated; BR, branched (*iso* or *anteiso*); CYCLO, cyclopropane substituted.

Fatty acid patterns, generated by GC, can be stored in a computer and analysed by the standard numerical techniques of ordination and cluster analysis, to provide classifications. Moreover patterns from unknown organisms can be identified by comparison with those stored in the computer.

Bacterial membranes are largely composed of amphipathic polar lipids, which comprise a hydrophilic head group linked to two hydrophobic fatty acid chains. Polar lipids are often referred to as free lipids, since they can be readily extracted by soaking cells in appropriate organic solvents. They are generally analysed by TLC, and often yield characteristic patterns. The most common polar lipids are phospholipids, other types include glycolipids and amino acid amides. Some polar lipids, for example, phosphatidylethanolamine, phosphatidylglycerol and diphosphatidylglycerol are common and of little diagnostic value, but others, such as the phosphatidylinositol mannosides of actinomycetes are rarer, and of more diagnostic importance (Table 3.4). A rapid presumptive test for archaebacteria is the absence of esterified lipids and the presence of diether lipids.

The isoprenoid or respiratory quinones are found in the plasma mem-

Table 3.4 Lipids of some actinomycetes with a wall chemotype IV (from Goodfellow and Cross, 1984).

Taxon	Long chain* fatty acids	No. of carbons	No. of double bonds	Predominant menaquinone†	Diagnostic phospholipids‡
Corynebacterium	SU(T)	22 – 38	0 – 2	MK-*8, 9*(H₂)	PI, PIM
Mycobacterium	SUT	60 – 90	1 – 2	MK-9(H₂)	
Nocardia	SUT	46 – 60	0 – 3	MK-*8*(H₄)	PE, PI, PIM
Saccharomonospora	SUIA	—	—	MK-*9*(H₄)	PE, PI, PIM

* S, straight chain; U, monounsaturated; T, tuberculostearic; I, *iso*; A, *anteiso*
† Abbreviations exemplified by: MK-*8*(H₂), menaquinone with two of the eight isoprene units hydrogenated
‡ PE, phosphatidyl-ethanolamine; PI, phosphatidylinositol; PiM, phosphatidyl-mannosides
Parentheses indicate not found on all strains

branes of all aerobic bacteria and comprise two types; the menaquinones (2-methyl-3-polyprenyl-1, 4-naphthoquinone, formerly known as vitamin K_2) and ubiquinones (2,3-dimethoxy-5-methyl-6-polyprenyl-1,4-benzo-quinone, or coenzyme Q). These are both large classes of molecules, in which the length of the polyprenyl side chain can vary from 1 to 14 isoprene units and also in the degree of saturation. They are generally extracted with organic solvent and analysed by reverse-phase TLC and/or high performance liquid chromatography (HPLC) (Collins, 1985).

Most bacteria contain either menaquinones or ubiquinones, or both. For example, the aerobic Gram-negative rods generally contain only ubiquinones with members of *Pseudomonas* possessing ubiquinones with 9 isoprene units, whereas several enteric genera contain mixtures of menaquinones and ubiquinones. Most members of *Bacillus* have menaquinones, and lactic acid bacteria generally lack isoprenoid quinones. Collins and Jones (1981) have surveyed the distribution of isoprenoid quinones amongst bacteria.

3.6 END-PRODUCTS OF METABOLISM

Traditional taxonomic tests, such as the methyl red test or Voges–Proskauer reaction, determine the end-products of metabolism, but for some taxa a more specific analysis of these metabolites is useful. Quantitative analysis using GC or HPLC is particularly valuable and lends itself to computerized data processing for generating classifications and effecting identifications.

Fermentative bacteria usually yield more complex patterns of end-products than aerobic strains, and the technique is particularly valuable for anaerobic bacteria. Most methods concentrate on the acid end-products, particularly the monocarboxylic acids such as acetic, propionic, butyric etc..

3.7 COMPLETE CELLS

When bacteria are grown under standardized conditions and thermally degraded in an inert atmosphere (pyrolysed, Py), the products provide a 'finger print' of the cell. These chemical profiles are very complex but can be analysed by either GC or mass spectrometry (MS). Most of the earlier analytical work used GC, but problems with reproducibility led to the development of Py – MS which is faster, more reproducible and more easily automated than Py – GC. The details of Py – MS have been described by Gutteridge *et al.* (1985); suffice to say that the output is a complex pattern of peaks reflecting the different products of pyrolysis. Since most bacteria tend to produce the same pattern of peaks, the quantitative data, represented by peak heights, which vary significantly and reproducibly between taxa, are used. The data are generally analysed using ordination procedures, since cluster analysis has not proved entirely successful, probably because of the complexity of the data. For this reason Py – MS is used to confirm existing classifications rather than generate new ones. It is ideally suited for identification purposes (see Chapter 6) since it is rapid (< 10 min per sample), widely applicable and readily automated. The only drawback is the cost (£80–100K)!.

3.8 CONCLUSIONS

Chemical analyses have become a major influence in bacterial taxonomy as new methods and techniques are developed. They have proved particularly successful for classifying and identifying organisms in which morphological and physiological characters have been few or have led to confusing classifications through undue importance being placed on them. This is particularly relevant to the actinomycetes and related Gram-positive bacteria, the classification of which has been revolutionized by chemosystematics (Goodfellow and Cross, 1984). However, it is considered by many that the principal contribution of chemosystematics derives from nucleic acid analyses largely because these are universal molecules, the composition of which is invariant with environmental change. DNA and rRNA sequence homology can therefore be used to provide a stable, comprehensive classification of bacteria and the ranking of taxa can be quantified to alleviate the over- and underclassification of genera, such as enteric genera and *Lactobacillus* or *Bacillus,* respectively. Finally, chemotaxonomic analysis, such as SDS-PAGE of cellular proteins, GC analysis of lipids and Py – MS, lend themselves to computerized data handling and automation which is having a tremendous influence in rapid, automated identification.

REFERENCES

Bradley, S. G. and Mordarski, M. (1976). Association of polydeoxyribonucleotides of deoxyribonucleic acids from nocardioform bacteria, in *Biology of the Nocardiae.* (M. Goodfellow, G. H. Brownell and J. A. Serrano, Eds.), pp. 310 – 336. Academic Press, London.

Brenner, D. J. (1970). Deoxyribonucleic acid divergence, in *Enterobacteriaceae Developments in Industrial Microbiology*, **11**, 139 – 153.

Britten, R. J. and Kohne, D. E. (1966). Nucleotide sequence repetition in DNA, *Carnegie Institute Yearbook* **65**, 78 – 106.

Collins, M. D. (1985). Isoprenoid quinone analysis in bacterial classification and identification, in *Chemical Methods in Bacterial Systematics* (M. Goodfellow and D. E. Minnikin, Eds.), pp. 267 – 288. Academic Press, London.

Collins, M. D. and Jones, D. (1981). Distribution of isoprenoid quinone structural types in bacteria and their taxonomic implications, *Microbiological Reviews*, **45**, 316 – 354.

De Ley, J., Segers, P. and Gillis, M. (1978). Intra- and intergeneric similarities of *Chromobacterium* and *Janthinobacterium* ribosomal ribonucleic acid cistrons, *International Journal of Systematic Bacteriology*, **28**, 154 – 168.

Dobson, G., Minnikin, D. E., Minnikin, S. M., Parlett, J. H., Goodfellow, M., Riddell, M. and Magnussum, M. (1985). Systematic analysis of complex mycobacterial lipids, in *Chemical Methods in Bacterial Systematics*, (M. Goodfellow and D. E. Minnikin, Eds.), pp. 237 – 267. Academic Press, London.

Gillis, M. and De Ley, J. (1980). Intra- and intergeneric similarities of ribosomal ribonucleic acid cistrons of *Acetobacter* and *Gluconobacter*. *International Journal of Systematic Bacteriology*, **30**, 7 – 27.

Goodfellow, M. and Cross, T. (1984). Classification, in *Biology of the Actinomycetes* (M. Goodfellow, M. Mordarski and S. T. Williams, Eds.), pp. 7 – 164. Academic Press, London.

Gutteridge, C. S., Vallis, L. and MacFie, M. J. H. (1985). Numerical methods in the classification of micro-organisms by pyrolysis mass spectrometry, in *Computer-assisted Bacterial Systematics* (M. Goodfellow, D. Jones and F. G. Priest, Eds.), pp. 389 – 401. Academic Press. London.

Jackman, P. J. H. (1985). Bacterial taxonomy based on electrophoretic whole-cell protein patterns, in *Chemical Methods in Bacterial Systematics* (M. Goodfellow and D. E. Minnikin, Eds.), pp. 115 – 130. Academic Press, London.

Jantzen, E. and Bryn, K. (1985). Whole-cell and lipopolysaccharide, fatty acids and sugars of Gram-negative bacteria, in *Chemical Methods in Bacterial Systematics* (M. Goodfellow and D. E. Minnikin, Eds.), pp. 145 – 172. Academic Press, London.

Kersters, K. (1985). Numerical methods in the classification of bacteria by protein electrophoresis, in *Computer-assisted Bacterial Systematics*

(M. Goodfellow, D. Jones and F. G. Priest, Eds.), pp. 337 – 365. Academic Press, London.

MacDonell, M. T. and Colwell, R. R. (1984). The nucleotide sequence of 5S ribosomal RNA from *Vibrio marinus*, *Microbiological Sciences*, **1**, 229 – 231.

Owen, R. J. and Pitcher, D. (1985). Current methods for estimating DNA base composition and levels of DNA-DNA hybridization, in *Chemical Methods in Bacterial Systematics* (M. Goodfellow and D. E. Minnikin, Eds.), pp. 67 – 93. Academic Press, London.

Pace, N. R., Stahl, D. A., Lane, D. J. and Olsen, G. J. (1985). Analyzing natural microbial populations by rRNA sequences, *American Society for Microbiology News*, **51**, 4 – 12.

Rogers, H. J. (1981). *Bacterial Cell Structure*. Van Nostrand Reinhold, Wokingham.

Stackebrandt, E. and Woese, C. R. (1981). The evolution of prokaryotes, *Symposium of the Society for General Microbiology*, **32**, 1 – 32.

Stackebrandt, E. and Woese, C. R. (1984). The phylogeny of prokaryotes, *Microbiological Sciences*, **1**, 117 – 122.

Staley, T. E. and Colwell, R. R. (1973). Application of molecular genetics and numerical taxonomy to the classification of bacteria, *Annual Review of Ecology and Systematics*, **4**, 273 – 300.

Williams, R. A. D. and Shah, H. N. (1980). Enzyme patterns in bacterial classification and identification, in *Microbiological Classification and Identification* (M. Goodfellow and R. G. Board, Eds.), pp. 229 – 338. Academic Press, London.

4 Phylogenetics

Phylogeny is probably not the best type of relationship with which to construct a classification (see Chapter 1), but inferring pathways of ancestry from phenetic information can be an interesting and rewarding pursuit. Recently, clearly defined procedures for deducing phylogenies have been developed and applied to microorganisms, but much remains to be done particularly with regard to analysis of macromolecular sequence data.

4.1 GENERAL CONSIDERATIONS

It is generally assumed that evolution follows a pattern of successive branchings into populations in which evolutionary change subsequently proceeds independently. The aim of the phylogeneticist is to determine the pattern of this branching and represent it as a tree (cladogram) from ancestral forms through to the organisms as we see them today. These cladograms may have a time axis, an evolutionary distance axis or may simply record the branching of lineages. This chapter will deal with methods for generating cladograms, but first it is necessary to cover two general aspects of this discipline.

4.1.1 Parsimony

The concept of parsimony dates from Darwin and postulates that evolution proceeds along the shortest possible pathway with the fewest number of steps. This concept of *minimal evolution* is essential if unrealistic theories of evolutionary change are to be avoided and is used to construct minimal length or most parsimonious cladograms. Sneath (1983) suggests a less restricted definition of parsimony as 'preferring fewer assumptions or simpler explanations', which then includes maximal compatibility cladograms. These are based on compatibility with the largest number of characters, irrespective of the number of changes that need to be made. Indeed, minimal length and maximal compatibility cladograms are closely related mathematically.

4.1.2 Trees

Cladograms are displayed as branched diagrams or trees. These trees may be rooted, that is have an origin that represents the common ancestor and from which the branches extend, or may be unrooted with no origin or direction (Fig. 4.1a). Whether the tree is rooted or not has considerable bearing on the computation involved. For four individuals, there are 15 different rooted trees but only 3 unrooted versions, and, for five individuals, the figures are 90 and 15, respectively (Fig. 4.1b). It is therefore usual to calculate an unrooted tree and choose the root such that distances to the OTUs are as equal as possible.

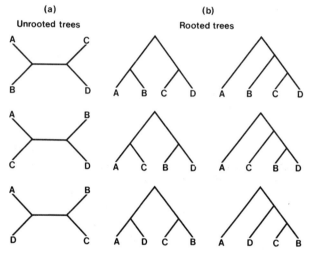

Fig. 4.1 (a) The three possible unrooted trees for four individuals. (b) Six of the 15 different rooted trees for four individuals.

The concept of maximal parsimony requires determination of the shortest possible tree. There are two types of minimum length (Fig. 4.2), depending on whether hypothetical ancestral organisms are introduced. Cladograms generally introduce these hypothetical taxonomic units (HTUs) at appropriate positions to produce shorter (Wagner) trees (see section 4.3).

4.2 DATA FOR PHYLOGENY

For many years, the only data available for inferring phylogeny were traditional phenotypic characters. For the animal and plant systematist,

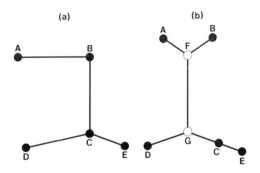

Fig. 4.2 (a) A shortest spanning (minimal length) tree. (b) A minimal length or Wagner tree introducing two new nodes (hypothetical ancestral organisms) at F and G (after Sneath, 1974).

careful choice of character together with a wealth of fossil and embryological evidence allowed the construction of plausible cladograms, and culminated in the works of Hennig and his theory of monophyletic groups (see Chapter 1). With bacteria, phenotypic characters in the form of morphology, cell structure and comparative biochemistry and physiology have been available for some time, but they were of limited use for phylogenetic purposes because the choice of early (primitive) and derived characters was entirely subjective. There was (and still is) no useful fossil evidence with which to confirm ancestral phenotypes. Thus, theories of bacterial cladogeny were numerous and almost invariably contradictory. The morphological scheme of Kluyver and Van Niel assumed that the coccus is the simplest and, therefore, most primitive form, and the more complex structures, such as rods and mycelia, are derived from this. One line of descent resulted in the streptomycetes, another in the endospore-forming bacilli, and a third in the spirilla. By contrast, Lwoff and Knight suggested that bacteria evolved with a progressive loss of anabolic activity. Thus, the nutritionally less exacting bacteria are the most primitive and the more fastidious have evolved a specialized way of life more recently. Conversely, it may be argued that fastidious heterotrophs living in the 'primordial' soup, from which all life forms were derived, were the most primitive. It is, therefore, apparent that phenotypic features are of limited value for deriving a cladogeny for bacteria, since there is no evidence to favour any one of the above theories and there is negligible information about any selective pressures that might have operated in natural habitats over long periods of time.

The possibility of constructing a phylogeny of microorganisms became more realistic with advances in molecular biology. If it is assumed that all extant life evolved from the same ancestor, traces of that ancestral organism should still be present in current species. Given a constant rate of evolutionary change, divergent evolutionary pathways and no gene transfer (but see Sections 4.3.1 and 4.3.2), homologous sequences in DNA, RNA or protein must represent inherited sequences from ancestral forms. The amount of homology will reflect the phylogenetic relatedness; high homology implies recent divergence from a common ancestor; low homology implies early divergence from a primitive ancestor. Therefore, current approaches to prokaryotic phylogeny use macromolecule sequences to construct cladograms consistent with theories of rates of evolutionary change, divergence and convergence of pathways and possibilities for lateral gene transfer.

4.3 PHYLOGENIES FROM MACROMOLECULAR SEQUENCES

The process of deriving a cladogram from macromolecular data is essentially the same, irrespective of whether DNA, RNA or protein sequences are used. In general, DNA is of limited value because rapid evolutionary change has left relatively few sequences in common between even closely related bacteria. The highly conserved function of ribosomes and some enzymes has limited divergence, and these molecules retain homologous sequences even between distantly related organisms. They are, therefore, the most useful for most phylogenetic purposes.

Working with rRNAs or proteins, the first stage is to align the sequences from different sources such that the maximum amount of homology is obtained. Usually, this will involve inserting gaps into sequences to allow for additions or deletions of residues. It is then necessary to calculate the differences between each sequence. The simplest procedure is to count the percentage differences between the two molecules, which is defined as the number of positions at which the two molecules differ per 100 residues. An approximately linear relationship to evolutionary change is observed for highly related molecules, but for more distantly related molecules some form of transformation is required to account for double and back mutations that are not detected by simple frequency counts. Various transformations based on theoretical and empirical premises have been proposed for protein sequences. The equation of Margoliash and Smith (1965) calculates the expected number of changes:

$$D' = n\ln[n/(n-D)]$$

where there are D differences in n sites. For rRNA sequences, Hori and Osawa (1979) calculate the rate of nucleotide substitution:

$$K_{nuc} = -(3/4)\ln[1.0-(4/3)\lambda]$$

where λ is the fraction of different residues. These corrections give a higher number of estimated changes than the observed changes.

Once the relationship between the molecules has been calculated, the next stage is the construction of the tree. There are two approaches to this. In the first, matrix methods use the table of sequence differences without recourse to the characters or character states. Such methods are useful for DNA reassociation data or serological analyses of proteins where actual sequences are not known. Such methods, for example UPGMA clustering (see Chapter 2), provide a phenogram and the result is heavily influenced by assumptions of evolutionary rates and divergence. If evolutionary rates are constant in different lineages and convergence is minimal the phenogram and cladogram will be congruent (the ultrametric property). Woese and his colleagues (Stackebrandt and Woese, 1982) have used UPGMA clustering of similarities of 16S rRNA catalogue sequences from prokaryotes extensively with impressive results (see Fig. 4.3e).

The second approach determines the shortest tree, and there are two methods for achieving this; namely statistical methods and those based on parsimony (although these also make statistical estimates). Overtly statistical methods are best represented by the maximum likelihood approach. Given three entities, the data, a possible evolutionary tree and a probabilistic method of evolutionary change, the probability of obtaining the data given the tree and model is computed. The tree that

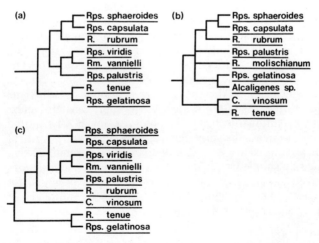

Fig. 4.3 Cladograms of purple photosynthetic bacteria derived from (a) cytochrome c_2 sequences (Woese *et al.*, 1980), (b) cytochrome c' (Ambler, 1985) and (c) 16S rRNA catalogues (Woese *et al.*, 1980). *Rps., Rhodopseudomonas; Rm., Rhodomicrobium; C., Chromatium.* (Adapted from Ambler, 1985).

maximizes this probability is the maximum likelihood estimate. It should be noted that this probability is *not* the probability that the tree is the correct one given the data and the model (Felsenstein, 1983).

Methods based on parsimony rather than a statistical analysis are more common, however, and involve determining the minimum length or Wagner tree. The length of a tree is the summed length of its branches, and algorithms for estimating Wagner trees from data matrices and for predicting the characteristics of the internodes (HTUs) are available. Under certain circumstances, the Wagner tree may be taken as a maximum likelihood estimate of the phylogeny and it provides the most parsimonious cladogram. Moreover, most parsimonious methods give better results than phenetic (matrix) methods if unequal evolution rates have occurred since they do not require evolution rates to be constant.

4.3.1   Constancy of evolutionary rates

Cladograms based on rRNA or protein sequences will fit the phenogram of the same data if evolution rates are constant in different lineages. However, if some lines evolve faster than others, then the deeper branching observed in some taxa will not reflect earlier divergence but a faster evolutionary 'clock'. To check if individuals within a group are isochronic (evolving at a similar rate), individuals within the group can be compared with an outgroup reference. If each member of the group shows a similar level of sequence homology with the reference, it may be argued that they have evolved at similar rates with regard to the reference. When Stackebrandt and Woese (1982) used this approach to evaluate evolution rates derived from rRNA catalogues of bacilli, they showed that most bacilli appear to evolve at a constant rate with respect to *E. coli*.

When protein sequences are used for inferring phylogenies, it must be remembered that different functional classes of protein evolve at different rates. Histones seem to be among the most slowly evolving proteins with one substitution occurring in several hundred million years. However immunoglobulins evolve about 100 times faster. The difference in evolutionary rates is largely due to functional constraints on the molecules; indispensable proteins with conserved function have a lower probability that an amino acid substitution would be compatible with maintenance of function. When a time scale is added to a cladogram it is therefore important to compare data from different molecules.

If the absolute time is known for a branching event it is possible to calculate the approximate times for other branches of the cladogram. Fossil evidence can be used to measure times of divergence, but the fossil record of early organisms (older than 100 million years) is poor. This, coupled with the uncertainty about rates of nucleotide substitutions in early organisms complicates adding a time scale to cladograms of bacteria.

5s rRNA sequence analysis suggests that Gram-positive and Gram-negative bacteria diverged about 1.2×10^9 years ago and geological evidence indicates that the photosynthetic bacteria were established $3.5 - 3.8 \times 10^9$ years ago, within a billion years of the earth's formation.

4.3.2 Convergent evolution and gene transfer

The prevalence of gene transfer in bacteria has led to the suggestion that the assumption of divergent branching in cladogram construction may not be acceptable and evolutionary lineages may frequently converge. Evidence for convergence through gene transfer can be obtained by searching for similar genes among organisms with very different backgrounds. Identification of such genes either represents extreme conservation or, more likely, recent transfer of the gene. Some antibiotic resistance genes are in this category, for example, a particular class of β-lactamases (TEM β-lactamases) are immunologically related and have considerable DNA homology, whatever this origin. This suggests that the gene for TEM β-lactamase has been transferred laterally amongst the bacterial kingdom. The later discovery that the gene was situated within a transposable DNA sequence (transposon) which could integrate and excise from bacterial, plasmid or phage chromosomes, and in so doing replicate itself, largely explained the distribution of this gene (see Hardy, 1981). More recently, several phenotypic traits have been associated with plasmids and often transposons, including hydrocarbon and sugar catabolism, pathogenicity in plants and animals, heavy metal tolerance, and, of course, antibiotic resistance (Hardy, 1981). There is overwhelming evidence that gene transfer is an important feature in prokaryotic evolution.

If gene transfer was a major feature of prokaryote evolution, however, it would be expected that cladograms based on different data would be incongruent since it would be unlikely that patterns of transfer for different genes would be the same. Although studies in this area are few, the phylogeny of the purple non-sulphur photosynthetic bacteria has been deduced from amino-acid sequences of cytochrome c and 16S rRNA cataloguing. The resultant cladograms were similar but not identical (Fig. 4.3), suggesting 'recent' transfer of cytochrome c genes between some of these bacteria (Ambler, 1985).

Perhaps, the most remarkable example of possible gene transfer in evolution concerns the enzyme superoxide dismutase. There are three forms of this enzyme; a copper – zinc-containing molecule found in eukaryotes, a manganese enzyme from prokaryotes and mitochondria, and an iron-containing enzyme restricted to prokaryotes. The iron- and manganese-containing enzymes are clearly related and originated from a common ancester, but the copper – zinc-containing enzyme is distinct. *Photobacterium leiognathi* is a symbiont that lives in a specialized gland in the pony fish and was thought to be unique amongst prokaryotes in possessing both iron- and copper – zinc-containing superoxide dismutases. The copper – zinc enzyme is closely related to the ponyfish superoxide dismutase, and one plausible explanation seemed to be that the gene responsible has been transferred from fish to bacterium! However, the recent discovery of this enzyme in some pseudomonads makes this supposition less likely.

In conclusion, the level of inter-species gene transfer is difficult to judge from the scant data available, but it seems likely that it will be sufficiently low to allow the formulation of reasonably accurate phylogenies of bacteria.

REFERENCES

Ambler, R. P. (1985). Protein sequencing and taxonomy, in *Computer Assisted Bacterial Taxonomy* (M. Goodfellow, D. Jones and F. G. Priest, Eds.), pp. 307 – 335. Academic Press, London.

Felsenstein, J. (1983). Methods for inferring phylogenies: a statistical view, in *Numerical Taxonomy* (J. Felsenstein, Ed.), pp. 315 – 334. Springer Verlag, Berlin.

Hardy, K. (1981). *Bacterial Plasmids*. Van Nostrand Reinhold, Wokingham.

Hori, H. and Osawa, S. (1979). Evolutionary change in 5S RNA secondary structure and a phylogenic tree of 54 5S RNA species, *Proceedings of the National Academy of Sciences of the United States of America*, **76**, 381 – 385.

Margoliash, E. and Smith, E. L. (1965). Structural and functional aspects of cytochrome *c* in relation to evolution, in *Evolving Genes and Proteins* (V. Bryson and H. J. Vogel, Eds.) pp. 221 – 242. Academic Press, New York.

Sneath, P. H. A. (1974). Phylogeny of micro-organisms, *Symposium of the Society for General Microbiology*, **24**, 1 – 39.

Sneath, P. H. A. (1983). Philosophy and method in biological classification, in *Numerical Taxonomy* (J. Felsenstein, Ed.), pp. 22 – 37, Springer Verlag, Berlin.

Stackebrandt, E. and Woese, C. R. (1982). The evolution of prokaryotes, *Symposium of the Society for General Microbiology*, **32**, 1 – 31.

Woese, C. R., Gibson, J. and Fox, G. E. (1980). Do genealogical patterns in purple photosynthetic bacteria reflect interspecific gene transfer? *Nature*, **283**, 212 – 214.

5 Nomenclature

What is in a name? This is perhaps rhetorical, but it is nevertheless a necessary question. Names are labels that should convey a message about the organism. To mention the surname of a human being will conjure up a mental image of a group of related individuals. The use of the Christian name identifies an individual within that group. Confusion results if the same Christian name is applied to two or more human beings of the same family group. Similarly within bacteriology, it would be impossible for communication if two taxa possess the same name. Names are not designed to characterize taxa, but represent a means for effective communication by labelling the entities. These labels are, of course, man-made, and constitute the procedure referred to as nomenclature. However, the naming of bacterial groups evokes the best, and sometimes worst, attributes of a scientist. The naming of new bacterial taxa is regarded by some scientists as a prestigious race to obtain immortality in the realms of biology. Where haste is evident, the outcome may be distastrous for reputations and bacterial taxonomy. The message is that it is often easier to allocate new names than do all the essential comparative work. However, it seems ironic that some of the noted bacteriologists are immortalized in the genus names of notorious pathogens. For example, Pasteur, Shiga and Yersin are remembered by the genera *Pasteurella*, *Shigella* and *Yersinia*, respectively.

So, the major function of nomenclature is communication. Consequently, it is important that names are readily pronounceable, sufficiently distinct from each other to avoid confusion, and stable. Bacteria may have common names, such as anthrax bacilli, and the more important binomial name *Bacillus anthracis*. This system owes its origin to the distinguished eighteenth century Swedish naturalist, Linnaeus. Thus species (considered here to mean isolates with a high degree of phenotypic and genotypic similarities) names comprise two words; the first representing the genus name, e.g., *Escherichia*, and the second consisting of the species identity ($\equiv$ specific epithet), e.g. *coli*. In this example, the binomial name is *Escherichia coli*. Convention dictates that the species name shall be italicized, as with the above example. These names are subjected to rules, as discussed in the International Code of

Nomenclature of Bacteria (Lapage *et al.*, 1975). A primary function is to ensure uniformity across international boundaries. After all, confusion must surely reign if the same organism may be associated with a multiplicity of names depending upon the whims of scientists from different countries.

Initially, the procedures of bacterial nomenclature copied the system used by either the Botanical or Zoological code. The special needs for bacterial nomenclature were initially mooted at the First International Congress of Microbiology in Paris during 1930. As a result, a Commission on Nomenclature and Taxonomy was formed, with a remit to investigate the problems and make recommendations to the Congress. The resultant resolutions highlighted the problems of bacterial nomenclature, but agreed that, wherever possible, the naming of microorganisms should follow the existing Botanical and Zoological codes. However, it was also recommended that a representative committee, which would include representatives designated by the Botanical Congress, should be established with a view to examining the subject of bacterial nomenclature. This committee was named the Nomenclature Committee of the International Society for Microbiology. Its duties appertained to resolving problems in, and defining criteria to be used in, bacterial nomenclature.

At the Second International Congress for Microbiology, in London during 1936, it was agreed to formulate a code for bacteriological nomenclature, which was discussed at the Third International Congress of Microbiology held in New York during 1939. This Congress approved both the development of a formal code, and the establishment of a Judicial Commission. The Commission's terms of reference included the issuing of formal nomenclatural 'Opinions' upon request; the development of recommendations for emendation of rules of nomenclature; the preparation of lists of the names of microbial genera, and the publication of the International Rules of Bacteriological Nomenclature. The Proposed Bacteriological Code of Nomenclature was considered at the Fourth International Microbiological Congress in Copenhagen, 1947. This was duly revised and published (in English) in the *Journal of Bacteriology* and the *Journal of General Microbiology*. Rapid progress ensued, and at the next Microbiological Congress (Rio de Janeiro, 1950), publication of an 'official' journal, the *International Bulletin of Bacteriological Nomenclature and Taxonomy*, was authorized. Thereafter, the Bulletin grew into the much respected *International Journal of Systematic Bacteriology* (IJSB). As for the code, emendations and revisions occurred until the present form, with its legalistic connotations, i.e., the International Code of Nomenclature of Bacteria, was published in 1975. Alterations may only be made by the International Committee of Systematic Bacteriology at one of its plenary sessions.

Essentially, the Code seeks to ensure the presence of stable, clear meaningful names, which are derived from Latin or Greek, or if necessary, latinized. Of course, the name, unique to one taxon, must be validly described in the scientific literature, with the designation of a reference (type) strain. 'Priority' is given to the first valid publication of a name. However for a name to be valid, it must be published in the IJSB. Moreover, names should not be changed without good taxonomic reasons. Thus, the Code should have a stabilizing effect upon nomenclature.

The Code considers taxonomic ranks from the level of subspecies up to and including class. These ranks include subspecies, species, subgenus, genus, subtribe, tribe, subfamily, family, suborder, order, subclass and class. The use of the term 'variety' is not encouraged, as this term is regarded as synonymous with subspecies. However, 'biovar', 'serovar' and 'pathovar' are accepted, although they are not covered by the Code. It should be emphasized that taxonomic ranking above family level is generally unclear in the case of bacteria.

The formation of names may be a difficult task, because consideration has to be given to grammar. Perhaps, the most straightforward aspect is the formulation of genus names, which appear to be constructed in an arbitrary fashion. With this case, the name may be derived from any word, although it is desirable that the name is descriptive or, at least, taken from the surname of a person associated with the organism, in particular, or bacteriology in general. Names should not be used which are already established in botanical or zoological taxonomy. Whatever the choice, the name is feminized, and always written in italics or underlined. The species name (or epithet) is similarly italicized, and, as a general rule, it may not be hyphenated, and has to agree in gender with the genus name. On no account shall a specific epithet be used more than once to describe two or more species within the same genus. The specific epithet should convey some meaningful message about the organisms, such as habitat or a distinctive characteristic. Alternatively, there may be a wish to name a species after a person, preferably the scientist connected with the isolation and/or characterization of the organism. This is permissible within the rules of the Code. Similar criteria apply to the delineation of subspecific names.

It may be observed that within the scientific literature, it is accepted convention to write the species name in full initially, e.g., *Pseudomonas aeruginosa*, but the genus name may be subsequently abbreviated to one letter followed by a full stop, e.g., *P. aeruginosa*. This is satisfactory providing that confusion does not result, such as may happen if reference is made to two or more genera beginning with the same letter, e.g., *Acinetobacter and Actinomyces*. With these cases, the genus name should be spelt out in full, or, abbreviated to a two or three letter code, such as *Aci.* and *Act.* for *Acinetobacter* and *Actinomyces*, respectively.

Above the rank of genus, the names should be latinized, and feminine in gender, and plural. Moreover, the hierarchy up to and including orders, is identified by means of characteristic suffixes. Thus the names of subtribe, tribe, subfamily, family, suborder and order have a stem derived from that of the type genus, and end in 'inae', 'eae', 'oideae' 'aceae', 'ineae' and 'ales', respectively (see Table 5.1). These names start with a capital letter, but there is some disagreement about whether or not there is a need for italics. Essentially, italics are used in the USA, but not always in Great Britain. Although it has been emphasized that the names of higher categories are derived from the type genus, there is at least one notable exception to the rule. This example is the family Enterobacteriaceae, for which the type genus is not *'Enterobacter'* but *Escherichia*. The valid publication of a new taxon is always accompanied by deposition in a culture collection of a reference strain. This is designated as the 'nomenclatural type' or simply the 'type strain'. There is a variety of terminology for the type strain.

Table 5.1 Formation of bacterial names up to and including Order (based on Lapage *et al.*, 1975)

Taxonomic rank	Suffix	Example
Order	-ales	Pseudomonadales
Suborder	-ineae	Pseudomonadineae
Family	-aceae	Pseudomonadaceae
Subfamily	-oideae	Pseudomonadoidae
Tribe	-eae	Pseudomonadeae
Subtribe	-inae	Pseudomonadinae
Genus	—	*Pseudomonas*
Subgenus	—	(not for *Pseudomonas*)
Species	—	*Pseudomonas fluorescens*
Subspecies	—	*Pseudomonas pseudoalcaligenes* subsp. *citrulli*
Biovar	—	*Pseudomonas fluorescens* biovar I
Pathovar	—	*Pseudomonas syringae* pathovar *tabaci*

'Holotype' is the type strain reported by the original author. However if the original author only described one strain (this practice is not tantamount to good taxonomic procedure) and did not specifically regard it as the holotype, then it is regarded as the 'monotype'. If a subsequent investigator designates one of the original author's strains as the type culture, then it is reported as a 'lectotype'. In the situation that the original cultures have been lost, another scientist may propose a type strain, providing that it corresponds closely to the original description. This is referred to as the 'proposed neotype', and, two years after its publication in the IJSB, it is regarded as the 'established neotype'. Of course, the situation may arise that, after a neotype strain has been proposed, the original cultures are re-discovered. In this case, the Judicial Commission should be informed so that a ruling may be made. So, type strains have been delineated for each *bona fide* species and subspecies, and type species and type genera for genera and families, respectively.

The code makes allowances for the inability to change and reject names. The former could be attributed to typographical mistakes in the original description. The latter may be voiced through an Opinion of the Judicial Commission; whereupon the name is regarded as *'nomen rejiciendum'*, and loses its standing in bacterial nomenclature. Conversely, a name may be conserved (*'nomen conservandum'*), in which case it must be used instead of other previous synonyms. In essence, names may be 'legitimate' or 'illegitimate' depending upon whether or not they are published in accordance with the Code. Changes may result from improvements in taxonomy. For example, a species may be transferred from one genus to another, in which case a new combination of names will result. This is referred to by the citation comb. nov. (an abbreviation for *'combinatio nova'*) which appears after the new name. Similar abbreviations, i.e., sp. nov. (*'species nova'*) and gen. nov. (*'genus novum'*) are used after new species or genus names. As an example, the causal agent of bacterial kidney disease in salmonid fish was elevated to new genus status, as *Renibacterium salmoninarum* gen. et sp. nov. The correct citation involves the names of the original authors and the date of the publication. This means that for the above example, the pathogen is referred to as *Renibacterium salmoninarum* Fryer and Saunders 1980.

The starting date for the priorities in bacterial nomenclature was given as January 1980. During this month, the *Approved Lists of Bacterial Names* was published in the IJSB. Therefore, the task of naming new taxa was made easier, insofar as reference needed to be made to this list and its subsequent supplements. Undoubtedly, the efforts described above have rationalized bacterial nomenclature, and, with universal acceptance of the Code, should radically improve the situation for bacteriology. However, some prospective authors (including eminent scientists) try to bypass the rules, and, occasionally at scientific meetings, new names will be announced to the unsuspecting audience. This may or may not be followed by definitive publications. Sometimes, these names rapidly become widely used. A topical example is the name of the organism associated with legionnaires disease, i.e., *Legionella pneumophila*.

REFERENCE

Lapage, S. P., Sneath, P. H. A., Lessel, E. F., Skerman, V. B. D., Seelinger, H. P. R. and Clark, W. A. (1975). *International Code of Nomenclature of Bacteria and Statutes of the International Committee on Systematic Bacteriology and Statues of the Bacteriology Section of the First International Association of Microbiological Sciences*. American Society for Microbiology, Washington, D.C.

6 Identification and diagnosis

The identification/diagnostic processes are the end result of taxonomy. Isolates may only be identified if classification has been a preceding step and the resultant taxa given names or codes. If the organism represents an undescribed taxon, then it cannot be identified. This may appear obvious, but it is a common fault of diagnosticians who attempt to identify the unclassified. Occasionally, novel organisms are encountered that are evidently distinct from, but nevertheless resemble, recognized taxa. The favourite ploy is to label the isolate as 'presumptive' or 'atypical'. Thus, presumptive coliforms may ferment lactose solely because of the presence of plasmid activity. Certain outbreaks of disease described as atypical pneumonia, were eventually attributed to a new taxon, *Legionella pneumophila*. The dilemmas facing the diagnostician need careful attention in order to ascertain the precise nature of the job.

Identification is a comparative process by which unknown organisms are examined, and compared to the known. The first stage is to obtain pure cultures of bacteria. Although this need has been emphasized in earlier chapters, it serves well to re-iterate here the absolute requirement for pure cultures. If mixed cultures are used, the results of any identification service will be quite meaningless. However, even experienced bacteriologists make the occasional mistake, especially if, for example, bacilli are well entrenched with actinomycetes. Cultures should be streaked and re-streaked for single colony isolation at least three times in order to ensure purity. Moreover, it is sound policy to check purity regularly by means of Gram-stained smears. These will highlight any obvious contamination. Thereafter, the pure culture is ready for the onslaught of the diagnostician.

Cowan (1974) discussed three views for identifying medical bacteria. As there is great truth in those statements, it is appropriate to consider them here, At one extreme, the unknown isolate can be examined for all possible tests. When the results are available, use is then made of standard texts, such as *Bergey's Manual of Systematic Bacteriology* (Krieg, 1984). Although this approach is time consuming and may be expensive in materials, it is envisaged that the battery of test results will permit the ready identification of the organism. In practice, some important diagnostic tests are usually forgotten, which leads to the necessity for more work and further time is wasted. This approach to identification is widely used, but it is not suited for the very busy diagnostic laboratory, where time and personnel are at a premium. Conversely, there is a step-wise approach in which diagnostic schemes are followed progressively. Thus answers for the first tests will allow the diagnostician to proceed with the next series of tests, until an answer is obtained. Again, this may be highly time-consuming. However, there should be many occasions during which the diagnostician has reasonable grounds to suspect the identity of an unknown organism. Therefore, specialist schemes suitable for identifying the category of organism suspected may be consulted. Such occasions might include recovery of cultures on selective media, or from well described pathological conditions. It would appear that use is made of the experiences, and perhaps intuition, of the diagnostician. Mistakes will occur, but rarely will the judgement be questioned.

So far, oblique reference has been made to the use of tests involved in identification processes. Many of these are classical bacteriological tests, such as the Voges Proskauer reaction, whereas others represent recent developments in microbiology, such as determining the presence of specific subcellular components. The results for such tests are compared with data for known organisms. The absolute requirement for carefully executed testing regimes should therefore be apparent; reliable and reproducible methods are necessary. The discussion of test error and reproducibility in Chapter 2 applies equally well here, insofar as error will inevitably lead to erroneous identification. However, the error will be considerably reduced if the tests are carried out properly. Any short cuts or bad techniques will reduce the effectiveness of the identification process. Attention to detail is a prerequisite of sound identification practice. This is an opportune moment to consider the nature of the tests to be used. These will hopefully be sensitive tests giving clearly defined reactions depending upon the ingenuity of the scientist who initially constructed the scheme. Considering that a reaction is dependent upon the nature of the methods used, it is important to mimic, as closely as possible, the precise methods used in the construction of the scheme. For example if motility was initially determined from wet preparations after 18 h incubation in nutrient broth, the diagnostician should use the same method. In this case, motility at 18 h could appear to be negative after 24 h. A wrong result would be recorded, which could contribute to mis-identification. This has importance for the differentiation of closely related motile and non-motile species, such as *Aeromonas hydrophila* and *A. salmonicida*.

6.1   DICHOTOMOUS KEYS

Dichotomous keys (diagnostic keys) were amongst the first forms of bacterial identification scheme. Indeed, they may be found in the earlier editions of *Bergey's Manual of Determinative Bacteriology*. With dichotomous keys, identification is achieved in a step-wise fashion. The diagnostician proceeds along progressively branching routes depending upon positive or negative reactions to tests. If the test is positive then route 'A' is followed but if a negative result is recorded then an alternative pathway is used (see Fig. 6.1) The main drawback is that an incorrect result, or an organism with an aberrant feature, will send the diagnostician along the wrong branch of the key, leading to mis-diagnosis. Largely for this reason, dichotomous keys have lost favour to diagnostic tables, and only a few keys have been proposed since 1970. Therefore, it must be concluded that the older keys lack the benefit of more recent developments in bacterial taxonomy. Moreover, they are usually based on unreliable classical tests.

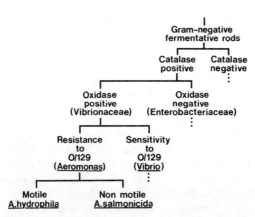

Fig. 6.1 Example of an identification key for *Aeromonas*.

6.2 DIAGNOSTIC TABLES

An end result of numerical taxonomy studies should be the construction of diagnostic tables (e.g., Table 6.1). These lack the disadvantages of dichotomous keys because they are polythetic and the presence of a few aberrant results is less likely to adversely influence the outcome of the identification process. With diagnostic tables, it is recognized that not all members of a taxon will give uniform test results. Thus, the presence of '+' or '−' tests indicate that most isolates

Table 6.1 Diagnostic table for identifying *Aeromonas*

Character	*Aeromonas hydrophila*	*Aeromonas media*	*Aeromonas salmonicida*
Gram-negative fermentive rods	+	+	+
Presence of brown diffusible pigment	−	+	+
Growth at 37 °C	+	+	−
Indole production	+	V	−
Degradation of casein	+	V	−
Motility	+	−	−
Phosphate production	+	−	V

+, − and V correspond to ≥ 80%, ≤ 20%, and 21 – 79% positive responses, respectively.

correspond to these results. Usually '+' and '−' correspond to ≥ 80 – 85% and ≤ 15 – 20% positive responses, respectively. Results indicated as 'V' or 'D' indicate variable positive responses, i.e., in the range of 21 – 79%. Diagnostic tables, which contain a matrix of test results for a range of bacterial taxa, may be found in widely used texts, such as Cowan (1974) and *Bergey's Manual of Systematic Bacteriology* (Krieg, 1984). The unknown isolate is examined for the tests indicated in the diagnostic table. Then the results are compared with the profile for each taxon. Obviously for large comprehensive tables, this may be a time-consuming process. However, there is scope for semi- and fully-automated procedures, such as offered by punch cards (Cowan, 1974) and computers, respectively.

6.3 COMPUTER-BASED IDENTIFICATION SYSTEMS

It is possible to use diagnostic tables in conjunction with computers. In the simplest form, the tables are incorporated into the computer memory, and the results for unknown organisms are compared with each of the possibilities and the highest correlation is taken as an identification.

Payne (1963) has been credited with the initial development of computer based identification procedures. Thereafter, the work of Dybowski and Franklin (1968) is noteworthy, insofar as these workers successfully identified medically important Gram-negative bacteria by use of computer techniques. In fact, a 50% success rate (6 out of 12 strains) was recorded. From this modest start, Lapage and co-workers from the Central Public Health Laboratory, Colindale, UK, successfully achieved reliable computer-based identification systems for 'hard-to-identify' Gram-negative bacteria of medical importance. Lapage *et al.* (1970) described a computer-based identification matrix comprising information for 62 reference taxa and their responses to 50 phenotypic tests commonly used in medical diagnostic laboratories. These included fermentation of 18 sugars and other characters known to be useful for fermentative Gram-negative bacteria, such as decarboxylase reactions, H_2S production, the methyl red and Voges Proskauer tests.

A data matrix was constructed in which each test result was converted into an estimate of the probability of a positive response. Thus values of 0.99 and 0.01 were allocated for results that were always positive and negative, respectively. Results that were *usually* positive or negative were scored as 0.95 and 0.05, respectively. Values of between 0.05 and 0.95 were attributed to variable test results depending on the frequency of positive results expected for the taxon. In these cases, the values were deduced after consultation of published literature and of Public Health Laboratory Service records. A computer program was devised based on Baye's theorem, which sought to identify isolates in three stages. These included the calculation of an 'Identification Score', determination as to whether or not a definitive identification was justified, and the selection of any further tests necessary to improve the chance for successful identification. Lapage and co-workers (1973) defined the likelihood of an unknown isolate belonging in a given taxon as 'the probability of obtaining the observed test results with a strain of this taxon. This probability is derived by multiplying together the probablities of the individual test results' (see Fig. 6.2). Thus, the computer compares the results of the unknown organisms against the entry for each taxon contained in the matrix. Where the unknown organism has a positive test result, the probability value is taken directly from the matrix. For example, if the unknown isolate produces catalase then during the comparison with a given taxon the probability of a positive result with this test is included in the calculation. However in the case of negative results for an unknown isolate, the probability is obtained by subtracting the probability of a positive response from unity (Fig. 6.2). The 'likelihood' figure gives an indication of which taxon the unknown isolate most closely resembles, but this may be more than one taxon and, therefore, a definitive identification is not obtained. Dybowski and Franklin (1968) approached this problem by calculating the percent relative likelihood in which the relative likelihood for each taxon is divided by the maximum likelihood obtained (Fig. 6.2). This is useful but still does not provide a critical level above which unequivocal identification can be assumed. Lapage *et al.* (1973) 'normalized' the identification score to provide such a level. By dividing the likelihood for each taxon by the sum of likelihoods for the matrix (normalizing), a figure is obtained that approaches 1.0 if the unknown identifies closely with one taxon and *very little with the others*. By choosing an identification score of 0.999, it can be safely assumed that

Determination of identification score
(After Lapage et al. 1970; 1973)

Data matrix:

Taxon	Probability of positive result for:		
	Catalase production	Oxidase production	β-Galactosidase production
1	0·80	0·95	0·05
2	0·99	0·01	0·95
Unknown isolate 'x'	+	−	+

Calculations:

Probability products ("likelihoods")
'x' compared with taxon 1: 0·80 × (1−0·95) × 0·05 = 0·002000
'x' compared with taxon 2: 0·99 × (1−0·01) × 0·95 = 0·931095
Sum = 0·933095

Percent relative likelihoods
'x' compared with taxon 1: 0·002000/0·931095 = 0·0021480085
'x' compared with taxon 2: 0·931095/0·931095 = 100%

Identification score (normalized)
'x' compared with taxon 1: 0·002000/0·933095 = 0·0021434
'x' compared with taxon 2: 0·931095/0·933095 = 0·9978566

Fig. 6.2 Determination of the identification score for bacteria (after Lapage *et al.*, 1970, 1973).

any isolate achieving this value must resemble only that one taxon. If it has characters in common with other taxa, the score will drop. Such a high identification score could be achieved with fermentative Gram-negative bacteria, but with less well described taxa, such as streptomycetes where overlap between taxa is more prevalent, an identification score of 0.99 is more realistic (Williams *et al.*, 1983).

The system, derived by Lapage *et al.* (1970), was a success insofar as 81.4% of the problematical Gram-negative organisms were identified correctly. When a limited number (i.e., 30) of tests was used the success rate was 77.4%. Some problems were uncovered, but these were resolved by modifying and extending the data matrix. Thus in a series of articles published in 1973, the matrix contained data for 51 tests on 70 reference taxa (Bascomb *et al.*, 1973; Lapage *et al.*, 1973). In these studies, 1079 reference cultures and 516 problematical fresh isolates of aerobic Gram-negative rods were examined against the data matrix. The outcome was that for fermentative organisms 90.8% of the reference strains and 89.4% of the fresh isolates were identified. Less success was recorded with non-fermenting organisms, insofar as identification was achieved with only 82.1% of the reference strains and 70.8% of the fresh isolates. Obviously, the future was bright for computer identification systems. Seemingly, the method could be applied to most groups of bacteria, providing that accurate data matrices could be established. Moreover, there was considerable potential for commercial exploitation.

6.4 SEROLOGY

Much confusion has resulted from the use of serology for bacterial identification, as may be deduced from the several hundred of *Salmonella* 'species' included in the seventh edition of *Bergey's Manual of Determinative Bacteriology*. Yet treated cautiously, a wealth of information may be derived from the reaction of antigen with antibody. Numerous serological reactions have been described, including whole-cell agglutination (WCA), latex agglutination, precipitin reactions, direct and indirect fluorescent antibody tests, and the enzyme-linked immunosorbent assay (ELISA) (see Smibert and Krieg, 1981). The main advantage is speed, and in some cases, such as with WCA, diagnosis may be achieved within a few minutes. With certain methods, e.g., ELISA, there is the potential for the development of kits suitable for field use, particularly directly on diseased tissue and without the need for isolating the offending organisms. In short, serology may be used to identify bacterial cultures, bacterial subcellular components, or bacteria embedded within tissues. The requirement is for reliable antibody (contained in antiserum).

In their crudest form, antibodies are raised in mammals, such as rabbits, rats, guinea pigs or mice, following injections with the antigen, i.e., whole bacterial cells or purified subcellular components. With time, the mammalian antibody producing cells (β-lymphocytes) secrete antibody into the blood, which may be removed, allowed to clot and thus give rise to antiserum. In the presence of sodium chloride, antisera react with antigens to give rise to a measurable reaction. If the reaction is homologous, i.e., the antibody associates with the antigen against which it was prepared, then the response is useful for diagnosis. However with heterologous reactions, the antibody cross-reacts with different antigens, and mis-diagnosis may ensue. The problem is associated with the complex antigenicity of the bacterial cell, and the presence of non-specific shared antigens. To some extent, heterologous reactions may be reduced by using antisera prepared against specific antigens, which are known to be restricted to the bacterial taxon under study. This requires a detailed knowledge of bacterial biochemistry, which is lacking for most taxa. Specificity of antisera may also be improved by development of mono-clonal antibodies, which will undoubtedly be used more extensively in the future. For the present, it is conceded that serology provides useful diagnostic information, which should not be used in isolation. Serological diagnosis of bacteria should be confirmed by other phenotypic traits. Nevertheless serology is of paramount importance in epidemiological studies for determining the serotype (or serovar) of any organism.

6.5 COMMERCIAL KITS

A cursory market survey would reveal the considerable potential for sales of identification kits, especially those aimed at medical laboratories. Indeed, many manufacturers have developed kits of which a few types have gained widespread use. However, most of these products have been designed for specific purposes, principally identifying certain groups of medically important bacteria. Within these constraints, it would appear that commendable success has been achieved, but problems may ensue when the kits are used for other purposes, such as for identifying a wide range of environmental isolates. In this context, it is important to note that the natural environment contains a wider range of taxa than that encountered in medical diagnostic laboratories.

Essentially two types of commercial kits have been developed; those that focus on biochemical responses of the organism, and those based on serological reactions. To date, the former have taken the larger share of an expanding market.

Commercially available kits measuring biochemical activity centre around rows of microtubes or paper strips impregnated with various freeze-dried test substrates. These are re-hydrated by inoculation with bacterial suspensions, and, after a pre-determined incubation period, the results are recorded as colour changes usually following addition of reagents. Identification may be achieved within 2 to 48 h. The simplest forms of kit, measuring a maximum of three biochemical tests, are marketed under the trade names of 'Dip slide' (Oxoid), 'Microstix' (Miles Laboratory) and 'Minitek' (Becton Dickinson). These kits were designed for Enterobacteriaceae, *Neisseria gonorrhoeae* and human pathogenic *Neisseria*, respectively. More comprehensive kits, which measure as many as 50 biochemical reactions, have been developed, and include the 'AP1' systems (AP1 Laboratory Products), 'Enterotube' and the 'Oxi-Ferm tube' (Roche), 'Minitek' (Becton Dickinson) and 'Patho-tec' or the 'Micro- 1 D' system (Warner Lambert). Variants of these products have been developed for identifying anaerobes, *Bacillus*, Enterobacteriaceae representatives, lactobacilli, pseudomonads, *Staphylococcus* and *Streptococcus*. With several of these systems, use may also be made of a computer-based data matrix for identifying unusual or atypical isolates. The oldest and one of the most widely used products is the AP1 – 20E rapid identification system, which was first marketed in 1969. This kit comprises 20 biochemical reactions, namely β-galactosidase, arginine dihydrolase, lysine and ornithine decarboxylases, utilization of Simmons citrate, H_2S production, urease, tryptophan deaminase, indole production, the Voges Proskauer reaction, gelatinase, and acid production from glucose, mannitol, inositol, sorbitol, rhamnose, sucrose, melibose, amygdalin and arabinose. In addition, the oxidase test needs to be recorded. The microtubes are re-hydrated with a few drops of the bacterial suspension, sterile mineral oil added to cover some tubes, incubated at 35 – 37 °C for 24 or 48 h and the reactions recorded, in some cases after the addition of reagents. Thus, there will be results for a total of 21 tests (including the oxidase test), which are coded numerically into a 7 digit code for comparison with a data matrix. The steps have been outlined in Fig. 6.3. It is our experience that the AP1 – 20E rapid identification system compares very favourably to the other commercially available products. A more rapid version, which permits identification within 4 h, has been marketed.

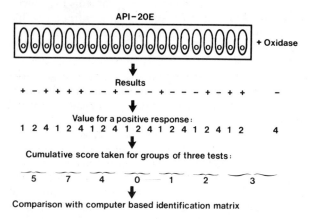

Fig. 6.3 Use of AP1-20E system for the rapid identification of bacteria.

Enterotubes differ from AP1 systems insofar as the former are inoculated by means of a wire that is inoculated with culture and then drawn lengthways through the compartments containing the already-hydrated media. Minitek comprises wells into which substrate containing discs are added. In contrast, Pathotec uses paper strips, which are impregnated with substrates. These are added to a bacterial suspension with incubation for 4 – 6 h. Some systems, e.g., AP1 – ZYM are based entirely on the detection of bacterial enzymes such as the ONPG reaction for β-galactosidase, a task which may be accomplished within 24 h.

The speed and ease with which these kits may be used is an obvious advantage. Thus the savings in media preparation and space compensate for the seemingly high cost of the kits. Nevertheless, there are problems with 'false' positive or negative activities. Of course, this criticism could also be made against the comparative conventional tests and in many cases the kits are more reliable. In the future, there may be attempts to interact commercial kits with identification 'machines'. Already pilot schemes appear to be very promising.

Serological tests are gaining popularity, particularly because of their ease and sensitivity. Kits have been developed for detecting a wide range of organisms, including *Bacteroides* spp. *Neisseria* spp., *Haemophilus influenzae*, and *Streptococcus* spp., by latex agglutination and immunofluorescence techniques. Moreover current research is aimed at developing and evaluating monoclonal antibodies destined for incorporation into the ultra-sensitive ELISA. Therefore, it is anticipated that the range of serological kits will be considerably extended.

6.6 BACTERIOPHAGE TYPING

The primary interest with bacteriophages is in epidemiological investigations. Thus the bacterial viruses are of great value in determining the presence of specific sub-groups ('types') of organisms. Indeed, bacteriophage typing schemes have been proven to be useful for many bacterial pathogens, including *Aeromonas salmonicida*, *Salmonella typhi* and *Staphylococcus aureus*. The technique involves use of virulent (lytic) bacteriophages which are inoculated by dropping onto freshly seeded bacterial lawns. With incubation for 18 – 48 h, a positive result is recorded by the presence of zones of clearing (plaques) in the otherwise uniform layer of bacterial growth. The reactions between bacteriophage and bacteria appear to be extremely specific. In practice bacteriophage usually lyse strains from the same taxospecies, occasionally from other species from within the same genus, and extremely rarely from strains of different genera. However, it is unclear how much emphasis should be placed on the few apparent cross-reactions.

6.7 CHEMOSYSTEMATIC METHODS IN IDENTIFICATION

Many chemosystematic methods are eminently suitable for identification purposes, particularly those that provide quantitative data which can be analysed by computer. Thus electrophoretic protein profiles, fatty acid profiles and Py – MS (see Chapter 3) are being developed for identification and have the advantages of speed and ease of automation over conventional tests.

Essentially, the approach involves the determination of profiles for reference strains, and either using these to prepare a classification or using an existing classification and storing the resultant data matrix in a computer. Profiles for unknown organisms may then be compared with patterns for reference taxa and identification is achieved if the patterns correlate. The data analysis is based on various discrimination procedures in which a combination of the quantitative characters that best differentiates an established set of taxa is determined. It is important to note that the taxa are not produced by these procedures (i.e., discriminant analysis does not classify the organisms); the groups must first be determined by hierarchical clustering or some ordination procedure. Discriminant analysis simply defines those characters that best distinguish the taxa.

One type of discriminant analysis is canonical variates analysis which is similar to principal components analysis (see Chapter 2). A transformed axis is determined which seeks to maximize the ratio of the variability between the means of the different taxa to that within the taxa. A second canonical variate is then sought orthogonal to the first and representing the next greatest variability, and subsequent axes are then derived. From this, a set of canonical variate means for each taxon can be determined and, for an unknown, a set of canonical variate scores is produced. For identification, the Euclidean distance of the unknown from each of the taxa can be determined. and the organism assigned to the closest (MacFie *et al.*, 1978).

Stepwise discriminant analysis may be used to determine those characters that maximize the ratio of variance between, to that within, groups. A subset of discriminant functions is calculated that provides a stable solution and these classification functions can be used to identify new samples. In the example shown in Fig 6.4 and Table 6.2, canonical variates analysis of Py – MS data shows that strains within the taxa *Bacillus amyloliquefaciens*, *B. pumilus*, *B. licheniformis* and *B. subtilis* could be discriminated according to this classification. Using stepwise discriminant analysis, 16 MS masses were needed to distinguish these groups effectively. Eight unknown strains were then analysed and cor-

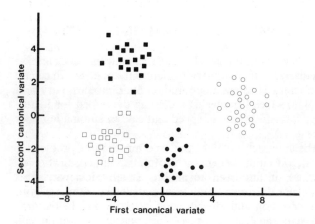

Fig. 6.4 Canonical variates analysis of four groups of *Bacillus* strains: ● *B. subtilis;* ○, '*B, amyloliquefaciens*'; ■ *B. pumilus*, □ *B. liqueniformis* (from Shute *et al.*, 1985, with permission).

Table 6.2   Identification of eight unknowns using stepwise discriminant (Taken from Shute *et al.*, 1985 with permission)

Actual identity	Distance from group mean				Identification
	S	P	L	A	
B.subtilis(S)	3.7	19.6	17.2	22.2	S
B. subtilis (S)	14.5	27.8	41.4	15.9	S
B. pumilus (P)	29.5	6.9	26.6	30.8	P
B. pumilus (P)	45.5	13.0	43.1	33.4	P
B. licheniformis (L)	24.6	24.4	7.2	38.4	L
B. licheniformis (L)	10.0	16.1	7.5	27.1	L
B.amyloliquefaciens(A)	39.8	20.8	32.5	11.6	A
B.amyloliquefaciens(A)	87.6	58.2	91.6	32.5	A

rectly identified (Shute *et al.*, 1985). Such powerful statistical procedures are now being applied to chemosystematic information, other than pyrolysis data, and the possibilities for automated rapid identification of microorganisms seem bright.

6.8   HYBRIDIZATION PROBES

The advent of recombinant DNA technology has provided a means of generating large amounts of specific DNA sequences (probes). These can be labelled, either radioactively with ^{32}P or using nucleotides derivatized with biotin so that they may be detected by autoradiography or by linking to an enzymic colourimetric system, respectively. These probes can thus be used to detect by hybridization, complementary DNA sequences which have been spotted on a nitrocellulose filter.

In the 'dot-blot' hybridization procedure, faecal samples, for example, are extracted with phenol to purify the nucleic acids present. They are then immobilized (dotted) onto a nitrocellulose filter in an array using a specialized vacuum (hybridot) manifold. The DNA samples are then hybridized with a labelled probe (e.g., ^{32}P) for a specific virus or toxin and hybrids are revealed by autoradiography. Positive samples are readily visible as dark spots.

DNA based identification is still in its infancy and the range of organisms to which it has been applied is small. However, it is increasingly popular in virology, and several diagnostic kits are now available for the detection and identification of RNA and DNA viral nucleic acids. Perhaps the best known bacterial system involves the use of toxin DNA sequences for the detection of enterotoxigenic *E. coli* in faeces. The advantages for DNA probes are specificity, rapidity and the ability to detect and identify the organism without having to isolate it from the source material. With the development of non-radioactive labelling of the probe DNA, the possibilities are certainly very exciting indeed.

REFERENCES

Bascomb, S., Lapage, S. P., Curtis, M. A. and Willcox, W. R. (1973). Identification of bacteria by computer: identification of reference strains, *Journal of General Microbiology*, **77**, 291 – 315.

Cowan, S. T. (1974), *Cowan and Steel's Manual for the Identification of Medical Bacteria*, 2nd edn. Cambridge University Press, Cambridge.

Dybowski, W. and Franklin, D. A. (1968). Conditional probability and the identification of bacteria: a pilot study, *Journal of General Microbiology*, **54**, 215 – 229.

Krieg, N. R. (Ed.). (1984). *Bergey's Manual of Systematic Bacteriology*, Vol. 1. Williams and Wilkins, Baltimore.

Lapage, S. P., Bascomb, S., Wilcox, W. R. and Curtis, M. A. (1970). Computer identification of bacteria, in *Automation, Mechanization and Data Handling in Microbiology* (A. Baillie, and R. J. Gilbert, Eds.), pp. 1 – 22. Academic Press, London.

Lapage, S.P., Bascomb, S., Willcox, W. R. and Curtis, M. A. (1973). Identification of bacteria by computer: general aspects and perspectives, *Journal of General Microbiology*, **77**, 273 – 290.

Macfie, H. J. H., Gutteridge, C. S. and Norris, J. R. (1978). Use of canonical variates analysis in differentiation of bacteria by pyrolysis gas-liquid chromatography, *Journal of General Microbiology*, **104**, 67 –74.

Payne, L. C. (1963). Towards medical automation, *World Medical Electronics*, **2**, 6 – 11.

Shute, L.A., Berkeley, R. C. W., Norris, J. R. and Gutteridge, C. S. (1985). Pyrolysis mass spectrometry in bacterial systematics, in *Chemical Methods in Bacterial Systematics* (M. Goodfellow and D. Minnikin Eds.) pp. 95 – 114. Academic Press, London.

Smibert, R. M. and Krieg, N. R. (1981). General Characterization, in *Manual of Methods for General Bacteriology* by (P. Gerhardt, Ed.), pp. 409 – 443. American Society for Microbiology, Washington, DC.

Sneath, P.H.A. (1978). Identification of microorganisms, in *Essays in Microbiology* (J. R. Norris and M. H. Richmond, Eds.), pp. 10/1 – 10/32. John Wiley and Sons, Chichester and New York.

Willcox, W. R., Lapage, S. P. and Holmes, B. (1980). A review of numerical methods in bacterial identification, *Antonie van Leeuwenhoek*, **46**, 233 – 299.

Williams, S. T., Goodfellow, M. Wellington, E. H. M., Vickers, J. C., Alderson, G., Sneath, P. H. A. Sackin, M. J. and Mortimer, A. M. (1983). A probability matrix for identification of some streptomycetes, *Journal of General Microbiology*, **129**, 1815 – 1830.

7 Interactions between taxonomy and allied disciplines

It is now relevant to consider some of the wider reasons for taxonomic studies. Of course, it is possible to argue the need for taxonomy from the philosophical standpoint, i.e., the acquisition of knowledge to satisfy the curiosity of taxonomists. This would have been an acceptable premise for the Victorian era when science was largely the hobby of the rich, but, alas, such freedom of action has long since waned. The current attitude is one of constant justification for resources. Therefore, science needs to have appeal in order to wrestle monies away from the ever-decreasing coffers. Fortunately, there are many areas of science in which a knowledge of bacterial taxonomy is essential or extremely useful. These include ecology, pathology, genetics and biotechnology, and culture collections.

7.1 ECOLOGY

The role of microorganisms in the natural environment has gained importance, particularly in relation to the survival of pathogens, nutrient cycling and pollution. There is a need to recognize certain components of a microflora, e.g., coliforms, that are indicators of unsanitary conditions. With the current popularity of biotechnology, there is interest in locating groups of organisms with special functions, such as antibiotic production. This necessitates detailed knowledge of the biology of the organisms. Unfortunately the classification of bacteria from the natural environment has been neglected (Table 7.1). Usually, ecologists have not been interested in taxonomy, and *vice versa*. Therefore, it is pertinent to determine the criteria for choosing meaningful representative strains from mixed microbial populations, such as occur in nature. Emphasis has been placed on general methods designed for medical bacteria rather than developing specialist systems suitable for ecology. Ecologists have identified isolates by means of conventional (medical) diagnostic schemes, although it is evident that the natural environment, including soil, leaf surfaces and water, is populated by bacteria that are likely to be different to those encountered in medical laboratories. Conversely, taxonomists,

Table 7.1. Some poorly defined or heterogeneous taxa associated with the natural environment.

Bacterial taxa	Leaf surfaces	Rhizosphere soil	Freshwater	Estuarine environment	Marine environment
				Present in	
Achromobacter spp.	+	+	+		+
Acinetobacter spp.	+	+	+	+	+
Aerobacter spp.	+		+		
Arthrobacter spp.		+			
Brevibacterium spp.	+	+			
Chromobacterium spp.	+	+	+	+	+
Corynebacterium spp.	+	+			
Flavobacterium spp.	+	+	+	+	+
Flexibacter spp.	+	+	+	+	+
Lactobacillus spp.	+	+			
Micrococcus spp.	+	+	+	+	+
Mycococcus spp.	+				
Paracolobactrum aerogenoides	+				
Pseudomonas carnea	+				
P. epiphytica	+				
P. ichthyodermis					
P. incognita	+				
P. schuylkilleonsis	+				
Sarcina spp.	+				
Staphylococcus spp.	+	+	+	+	+

who expressed interest in the species composition of natural microbial populations, have tended to examine isolates of dubious ecological significance. The end result could be chaotic, with many environmental isolates identified as poorly defined or heterogeneous taxa. Even with the use of modern procedures, such as numerical taxonomy, progress has been hampered by the extreme difficulty of identifying phena. Again, scientists have resorted to use of conventional schemes such as Cowan's (1974) identification tables or the older editions of *Bergey's Manual of Determinative Bacteriology*. With these, several distinct phena may be identified to the same species on the basis of a few 'key' features. Thus, these methods have been largely unsuccessful in identifying the wide variety of bacteria occurring in natural habitats. It is apparent that ' . . . the identification of an unknown organism presupposes that others like it have been previously examined, compared with more organisms and named' (Gray, 1969).

If the taxonomy is inadequate, it is very difficult to make meaningful judgements about the patterns in microfloras. Ecologists have tended to emphasize a few subjectively chosen 'key' morphological characteristics of microfloras. Colony pigmentation is readily observable trait; and this has been repeatedly highlighted. Thus, populations of leaf surface (phylloplane) and marine bacteria contain large proportions of yellow-pigmented types. In contrast, the soil contains a microflora that produces mostly cream or white colonies. It is tempting to infer that the phylloplane and marine microfloras are more closely related to each other than to the soil. Yet, many of the yellow-pigmented bacteria have been traditionally identified as *Flavobacterium*, which has become a heterogeneous taxon based largely upon pigmentation. Consequently, with reports that the phylloplane and the sea contain large populations of '*Flavobacterium*', it is easy to realize why scientists may link the two vastly different habitats. With further study, it has been determined that marine microorganisms comprise a diverse range of unique taxa. However in the coastal environment, a high proportion of the organisms may be derived from the land, constituting a 'wash-in' component. Indeed, the term 'resident' has been coined to distinguish the indigenous component of the microflora, capable of multiplying in the natural environment, from organisms that only have a transient phase. However from the techniques used to study microbial populations in the natural environment, it is difficult to distinguish between the resident and transient components.

Similarly, the soil contains some yellow flavobacteria, chromogenically related to a particular component of the phylloplane microflora. It is speculative whether or not the leaf surface bacteria originated in the soil, or *vice versa*. However, it is now realized that seed-borne bacteria may colonize both the phylloplane and rhizoplane (root surface); the latter being in intimate association with the soil.

Sweeping ecological generalizations have followed the acquisition of minimal data. For example, it was considered that many of the bacteria isolated from the phylloplane occurred on a wide range of plant species (Last and Deighton, 1965). Yet, it was unwise to put too much weight on distribution patterns because of the unsatisfactory state of classification of phylloplane bacteria. Nevertheless detailed investigation showed that some bacteria, notably *Beijerinckia and Spirillum* spp., were isolated only from tropical plants and appear to have restricted host ranges. The success of this study by Ruinen (1961) stemmed from the detailed examination of a few taxa, noted for their nitrogen metabolism, rather than a broad spectrum approach, by which sweeping generalization are made about all aspects of a natural microbial population. The latter may be unconvincing, even with more modern approaches.

It may be envisaged that the concept of numerical taxonomy led to welcome advances in ecology. Certainly, this technique has been used with bacterial populations derived from freshwater, estuarine and marine samples, soil, root surfaces, leaves and leaf litter. The first application was published by Brisbane and Rovira (1961), who studied 43 'rhizosphere' (the zone around roots) isolates and 21 reference cultures for 20 micromorphological, biochemical, degradative, antibiotic sensitivity and ecological tests. The study can hardly be considered as exhaustive, and the outcome of the analysis was not particularly helpful. Thus 31 of the environmental isolates were recovered in one cluster, defined at the 85% similarity (S) level, whereas the reference cultures clustered separately. This problem has re-occurred in many subsequent numerical taxonomy studies, involving environmental isolates, with reference cultures contributing little to the identification of phena. Part of the problem may reflect the difficulties with choosing meaningful reference strains, but there is the additional possibility that organisms, maintained in artificial laboratory conditions, lose some of the characteristics associated with freshly isolated bacteria. Nevertheless, after this auspicious start by Brisbane and Rovira, the numerical studies continued using more isolates and larger batteries of tests. Unfortunately, many of these studies have not capitalized on the outcome of the analyses to aid the understanding of ecology. It would appear that in many cases large collections of strains have been characterized solely for the purpose of carrying out numerical taxonomy studies. When taxonomy and ecology have interacted the outcome has been helpful. For example, the ecological study of soil bacteria by Lowe and Gray (1973 a, b) was only possible because of the initial taxonomy (Lowe and Gray, 1972). Lowe and Gray examined 209 isolates and 21 references cultures for 179 unit characters. Following analyses by the simple matching and Jaccard coefficients with clustering by single linkage, 180 of the isolates together with 8 marker strains were recovered in 7 clusters on the basis of overall similarity. Representative cultures were subsequently used in growth and competitive interaction experiments. In a larger study, Hissett and Gray (1974) examined 400 cultures, isolated from woodland litter and soil horizons, using numerical analyses. Although it is difficult from their publication to determine the precise techniques used, the end result showed that of 19 clusters, 7 phena contained soil isolates, 7 comprised litter isolates, and 5 phena contained both litter and soil organisms. Unfortunately the phena were not identified, but were somehow allocated to 'approximate taxonomic positions'. The study enabled soil isolates to be generally distinguished from those recovered from litter, and underlined differences between populations in these contrasting habitats. One of us (B.A.) used numerical taxonomy methods to study 621 isolates recovered from the phylloplane of perennial ryegrass. Again, difficulty ensued with attempts to equate phena with recognized taxa. However, representative strains were carefully chosen for ecological work, aimed at studying antagonistic interactions between phylloplane bacteria and a plant pathogenic fungus. The most convincing antagonist was *Pseudomonas fluorescens*, a taxon that appeared on the leaves only during the warmer summer months when the fungal pathogen, i.e., *Drechslera dictyoides* was most troublesome.

It may be concluded that numerical taxonomy has successfully enabled bacteria from heterogeneous populations to be grouped together in homogeneous clusters on the basis of shared characters. Once these clusters have been defined then *a posteriori* weighting of the group characters for use in diagnostic schemes becomes a valid proposition. However only a few of the studies, involving isolates of ecological importance, have attempted to devise diagnostic aids suitable for identifying fresh isolates. This is a serious shortcoming of the other investigations insofar as comparisons between studies is therefore more difficult to achieve. Other shortcomings in common with many of the numerical taxonomy studies include problems with the choice of stable reference cultures, the selection of 'good' characters, and the difficulties with negative correlations.

The inclusion of suitable reference cultures is at times a seemingly impossible task, and in the studies outlined above few were actually recovered in the clusters. For example, Lowe and Gray (1972) reported that only 38% of the named strains clustered with the environmental isolates. Even this low proportion is good compared to many other studies. Apart from the choice of suitable reference cultures, the selection of adequate characters has been a serious weakness of numerical taxonomy studies. The nature and number of the tests is subject to the discretion of each worker, yet these tests constitute a vital aspect of any investigation. Ideally a wide range of characters should be used from which an understanding of the overall properties of the organism may be achieved. It could be argued that ecologically based tests would be advantageous when studying organisms recovered from the natural environment. However, a danger would then exist that the work could not be compared to studies of other habitats. Consequently, it would be impossible to deduce whether or not the microflora of habitat 'A' was distinct from habitat 'B'. It is better, therefore, to use tests that can be directly applied or modified to suit the study of organisms from a wide range of habitats. The problems appertaining to test error and reproducibility have been discussed (see Chapter 2), yet numerical taxonomy studies of environmental isolates in many cases were based on small numbers of characters (e.g., Brisbane and Rovira, 1961), and errors could have caused severe distortions in the overall clustering of the isolates. Moreover, negative correlation (see Chapter 2) is particularly troublesome, insofar as many environmental isolates, e.g., *Flexibacter* spp., are quite unreactive in conventional testing regimes, and consequently could be clustered together on the basis of characters they did not possess. Thus, a false impression of homogeneity between dissimilar organisms would occur.

An aspect, for which large scale taxonomy studies has been useful, concerns the interpretation of the term 'species'. All too often, it is difficult to distinguish where one taxon finishes and the next begins. This problem has been observed in numerical taxonomy studies of Gram-negative bacteria recovered from the natural environment. Of course, species are man-made devices, and there is no sound reason why microorganisms should fit into convenient boxes. Both dendrograms and similarity matrices reveal areas of high similarity between some OTUs that are surrounded by strains of lesser similarity.

This situation could be reasoned as an example of convergent evolution in the natural environment, where dissimilar 'species' are effectively approaching each other in terms of phenotypic expression due to the effects of common pressures in the environment. It is also possible to prepare a case for divergent evolution to explain this situation. Nevertheless, it should be borne in mind that bacteria are capable of multiplying at an extremely rapid rate. Therefore, a change in a single phenotypic expression every million generations would soon be observed in a bacterial population. If such events occur at fairly regular intervals, the results would be marked heterogeneity among populations of related bacteria. Within clearly defined taxa, such as *Pseudomonas fluorescens*, there may be a diversity of phenotypic expression. A numerical taxonomy study revealed that there was variation in the response of 131 isolates to 41/136 test. Of course, the reasons for this variation are likely to be complex, but may include the effect of mutagens, e.g., ultraviolet light and chemicals on the bacterial DNA, genetic exchange between bacterial cells, i.e., by conjugation, transduction, transformation, or by plasmid transfer, progressive modification of the nucleoid, and the effect of the environment on phenotypic expression, namely temperature, availability of nutrients, and competition for space.

Taxonomy and ecology are delicately intwined, such that disregard for the former will adversely effect the latter. The moral is that good ecology can only result from sound taxonomy. The absence of sound taxonomic procedures may lead to highly subjective decisions about ecology.

7.2 PATHOLOGY

The role of microorganisms in disease processes was one of the principal founding interests in microbiology. It is apt to recall Koch's postulates, which essentially considered the need to isolate a pathogen in pure culture, to confirm its pathogenicity, and then to re-isolate the same organism from the infection. The first and last stages demand the need to recognize an organism. Yet 100 years after Koch, the ability of diagnosticians to identify organisms remains questionable. Perhaps, the problem concerns the large volume of work handled by most diagnostic laboratories, and, therefore, the need to spend only the minimal amount of time and resources on each specimen. This situation is complicated by the overwhelming need to kill the pathogen in the infected host, rather than to carry out extensive characterizations. Consequently, much attention is focused on antibiotic sensitivity testing. The outcome is that new or interesting pathogens may be missed. Moreover, epidemiology studies may be hampered. The taxonomy of pathogens receives effort only when new patterns of disease become apparent. The case of legionnaire's disease became well documented in the popular press after an oubreak in the USA, which resulted in numerous deaths. The causal agent was eventually isolated, and described in a new genus, as *Legionella pneumophila*. Subsequent investigation has revealed that legionnaire's disease is not a new condition, but has been regarded in the past as 'atypical pneumonia'. Even after considerable publicity, legionnaire's disease was initially mis-diagnosed as influenza (type B) during a serious outbreak in Stafford, UK, that occurred in spring, 1985. Similarly, *Erwinia herbicola* was announced as an emerging pathogen of human beings, yet a culture had been deposited in a culture collection 40 years previously, albeit mis-named as *Chromobacterium typhiflavum*. Therefore, it was necessary for good observation to pinpoint the presence of a hitherto undescribed human pathogen. Even *Aeromonas hydrophila*, an ubiquitous freshwater inhabitant, has been recognized to be a serious pathogen. Previously, isolates had been merely labelled as atypical coliforms. With the general awareness of the presence of these serious pathogens, meaningful epidemiological investigations have been possible.

The saga of unrecognized pathogens is not confined to human medicine. The taxonomy of plant pathogenic xanthomonads was complicated by the system of naming new species after the host plant, without detailed characterization to confirm the unique position of the so-called 'new' organisms. Fortunately, detailed taxonomic work has led to the recognition of synonyms, and thus to a marked reduction in the total number of *Xanthomonas* species. For example, one of us (B.A.) recognized the high level of relatedness between reference cultures of *X. begoniae*, *X. campestris* and *X. hedera*, which did not warrant the presence of three separate species names. Indeed, this group of xanthomonads was subsequently determined to comprise a dominant component of the phylloplane microflora of healthy plants. Therefore, this habitat would form the likely reservoir for infection.

Veterinary medicine has also benefitted from the attention of bacterial taxonomists. The speciation of coryneforms and mycobacteria has almost become a separate specialist branch of science. *Corynebacterium pyogenes* has been reclassified into the genus *Actinomyces*, as *Actinomyces pyogenes* largely on the basis of chemotaxonomy. Mycobacteriosis may appear to be a syndrome caused by a multiplicity of species, yet many of these may be reduced to synonyms of a few well defined taxa. Perhaps the confusion has been caused by an abundance of reports based largely on histological examination of infected material. In fact, some diagnosticians have not bothered to isolate the offending pathogens; instead, diagnoses have resulted solely from the presence of acid-fast organisms in tissue samples. It is relevant in these cases how mycobacteria may be differentiated from nocardia, which also cause serious infections in animals. Maybe, the diagnosticians are gifted with considerable taxonomic intuition! Yet, the all-important control regimes may differ from mycobacteria and nocardia. Therefore, mis-diagnosis may lead to the application of inappropriate treatment.

The blinkered approach to diagnosis often causes interesting pathogens to be missed. For example, *Vibrio anguillarum* has long been regarded as the causal agent of vibriosis in marine fish. The disease may be described as a haemorrhagic septicaemia. Consequently, many such cases of disease have been attributed to vibriosis, particularly if the pathogen displays certain key characteristics, namely growth on thiosulphate citrate bile salt sucrose agar, and the presence of motile Gram-negative fermentative rods which produce catalase and oxidase. Many organisms were identified as *V. anguillarum*, before heterogeneity among the isolates became obvious. Initially, the taxon was divided into two biotypes, of which the latter became reclassified as *V. ordalii*. The relevance of this observation to fish pathology is that *bona fide* strains of *V. anguillarum* are more serious as pathogens than *V. ordalii*. Moreover, this information has to be considered in the development of vaccines, which may necessitate the need for multivalent products. Other species of *Vibrio* are now recognized as causal agents of fish diseases, including *V. alginolyticus*, *V. carchariae*, *V. cholerae*, *V. damsela* and *V. vulnificus*. Undoubtedly as more attention is given to this area of pathology, other 'new' pathogens will be recognized.

Sometimes, impressions may be gained of serious epidemics affecting wide geographical areas. In one such case, in fish, streptococcicosis was identified in Japan, followed by outbreaks in South Africa and USA. However, the causal agent was not fully characterized; consequently, it is unclear whether the outbreaks were caused by one or more species. This is of paramount importance from the epidemiological viewpoint.

The difficulties, discussed above, have resulted largely from incomplete taxonomic knowledge. At one extreme, there is haste to describe new taxa. Conversely, poor taxonomy may lead to the failure to recognize patterns in disease outbreaks. There is a tendency to follow certain lines of work, regardless of whether or not they are suited to the conditions. Rapid diagnosis of disease is essential for swift corrective action. Serology is particularly suited for such speed, yet the specificity of the reactions has rarely been considered. It is appropriate to consider the hundreds of *Salmonella* 'species' included in the earlier editions of *Bergey's Manual of Determinative Bacteriology*. These 'species' were created solely as a result of serology. Fortunately, these taxa have been reduced in status, to serological variants of a few well defined species. Such finely divided groups are of value only for epidemiology, where it is important to trace the source and spread of infections. Thus, the ability to recognize specific (perhaps minority) components of large mixed populations is invaluable.

7.3   GENETICS AND MOLECULAR BIOLOGY

Until recently, the interaction between molecular biolgy and taxonomy has been entirely unidirectional, involving the adoption of molecular and chemical methods by taxonomists to improve their classifications and provide superior identification systems. Taxonomy has been largely ignored by molecular geneticists. This could be interpreted as flexibility of the systematist, who is prepared to explore new techniques, and the relatively narrow outlook of the molecular geneticist, who generally aims to analyse one particular strain in detail. This was understandable during the development of bacterial genetics, since exploitable systems of gene exchange were only available for a few organisms and even today genetic linkage maps of any detail have only been prepared for 15 bacteria (O'Brien, 1984). However, the situation is now changing, and the molecular geneticist might benefit greatly from reading some of the new systematics. The principal reason for this is the development of gene cloning techniques. It is now possible, theoretically at least, to introduce genes from a Gram-negative bacterium into any other Gram-negative bacterium and maintain them stably. This is possible through the use of the broad-host range plasmids of incompatibility group P as vectors for cloned genes (see Glover, 1984). For Gram-positive bacteria such vectors are not available and plasmids with more restricted host-ranges are used. In the case of *Bacillus subtilis*, a genetically well characterized organism, indigenous plasmids with selectable markers such as antibiotic resistance initially proved difficult to detect, and it was not until 1977 that plasmids from *Staphylococcus aureus* were found to be suitable for cloning in *Bacillus* owing to their stable replication and maintenance in this host. Had the close molecular relationship between these Gram-positive cocci and bacilli, which has now been demonstrated by rRNA cataloguing, been known earlier, it is likely that the *S. aureus* plasmids would have been exploited in *B. subtilis* genetics at an earlier stage. Thus, a knowledge of the classification of bacteria can indicate to the geneticist groups of closely related bacteria amongst which plasmids and phage might be exchanged and stably maintained.

In a similar vein, there are barriers to heterologous gene expression. Genes from Gram-positive bacteria are usually transcribed and translated in *E. coli*, but Gram-negative genes are not normally expressed in *B. subtilis* and other Gram-positive eubacteria. Streptomycetes seem to express only actinomycete genes. Again, a sound knowledge of the molecular relationships between bacteria should indicate hosts for gene cloning that readily express heterologous genes and can be used in particular situations, such as lactobacilli in silage production or pseudomonads for hydrocarbon degradation. It seems likely, therefore, that as molecular biology is applied to more and varied bacteria a comprehensive classification of these bacteria will be an invaluable aid to the molecular biologists developing the genetic systems.

7.4 BIOTECHNOLOGY

Biotechnology, the commercial exploitation of biology, relies heavily on microbiology and the microbial systematist is finding his skills are increasingly sought. The demand for novel pharmaceutical agents be they antibiotics, antitumour or antiviral agents, enzymes or enzyme inhibitors is great. The list of valuable primary and secondary metabolites is now very long indeed and continues to grow as novel bacteria are isolated and cultured. However a major limitation in the development of these technologies is the isolation of novel bacteria from the environment. Initially, microbiologists sampled exotic habitats ranging from insect intestines to the cold seas of Antarctica in the search for unusual bacteria and this was reasonably effective. However, if the normal media and incubation conditions are used, common organisms, such as endospore forming rods, enteric bacteria and pseudomonads fluorish during the isolation procedures and overgrow the rarer organisms. Even if the source material was derived from some exotic habitat it is likely that common organisms will be isolated. More recently, data from taxonomic studies have been used to approach the isolation of novel bacteria in a more scientific way (Williams *et al.*, 1984). In the initial study, an exhaustive numerical taxonomic analysis of *Streptomyces* and related organisms provided a comprehensive data matrix comprising many poorly represented and uncommon taxa as well as the more common ones, together with their characteristics. From this information, selective media were formulated, often containing three or more antibiotics against which strains of the selected taxon were known to be resistant, or comprising an unusual carbon source that could be used only by the selected organisms. Media were thus designed that would repress the more common organisms but permit the growth of selected unusual taxa. In this way, Williams and his colleagues were able to sample common habitats and isolate rare and unusual streptomycetes that were subsequently screened for the synthesis of unusual antibiotics. A second sample relates to the isolation and identification of insect pathogenic bacteria with potential as biological control agents. Some *Bacillus sphaericus* strains are toxic to mosquito larvae, but the distribution of toxicity amongst strains of this rather diffuse species was until recently unclear. DNA reassociation studies of 62 *B. sphaericus* strains revealed at least four homology groups, with all the mosquito pathogens assigned to group IIA. Phenotypic characters that distinguished this taxon were later identified, and it was possible to develop a medium for the selective isolation of strains belonging to Group IIA. Thus, many more strains with increased pathogenicity or different host ranges could be readily isolated.

Microbial taxonomists also contribute to the efficient operation of traditional biotechnological industries, such as the food and dairy industries, brewing and other industrial fermentations for enzymes or antibiotics. Any large-scale fermentation or manufacture of a food is susceptible to contamination during production or spoilage of a packaged product. Such contamination may be detrimental to the process and reduce yields, may affect the flavour, efficiency or appearance of the product or may be deleterious to health, if consumed. Rapid detection and identification of spoilage organisms is, therefore, vitally important to the efficient manufacture of these products, and although rapid detection methods are outside the scope of this book, computerized identification procedures (as outlined in Chapter 6) are proving to be increasingly useful. A typical problem could involve the microbial spoilage of beer. A recent numerical taxonomic study of Gram-positive cocci in beer and breweries revealed *Pediococcus damnosus* and *P. pentosaceus* strains, various staphylococci and micrococci (Lawrence and Priest, 1981). However, of these bacteria only *P. damnosus* strains are able to grow in and spoil beer; the low pH, and presence of hop α-acids inhibit the other bacteria, which presumably gain access to the beer from the air, raw materials and human contact. Isolation of staphylococci or micrococci from beer during normal quality control procedures could lead to panic decisions to pasteurize and recycle the beer or even destroy it if the bacteria were thought to be the spoilage organism *P. damnosus*. However, by operating a rapid identification scheme in the laboratory on a microcomputer, the quality control scientist can quickly and accurately identify the offending bacteria, and if they are harmless micrococci, which are unable to grow in and spoil the product, the beer can be processed as normal. Obviously the source of the bacteria would have to be located and eliminated, but there would be no immediate threat to the product. This is just one of many situations in which a rapid, reliable and predictive identification service can save substantial sums of money and lead to more efficient and productive manufacturing processes.

7.5 CULTURE COLLECTIONS

Culture collections are the microbiologists botanical or zoological garden, in which representative microorganisms are preserved and maintained. They provide an indispensable service to microbiologists through the provision of well characterized, authenticated strains of viruses, bacteria, fungi, protozoa and, more recently animal and cell lines. Many countries have, or are introducing, these 'service' collections, which supply microorganisms to universities, schools, hospital, research centres and industry. Some countries, such as the USA and West Germany, have single centralized culture collections (American Type Culture Collection and Deutsche Sammlung von Microorganismen, respectively), but in the UK ten specialized national culture collections hold nearly 30 000 strains ranging from viruses to animal cell lines. These collections are co-ordinated by the UK Federation for Culture Collections and are represented on the World Federation of Culture Collections, which provides international coordination.

In addition to the distribution of well characterized authenticated strains, culture collections are also centres of expertise in the fields of preservation, classification and identification of bacteria. Many culture collections offer comprehensive catalogues with detailed histories and applications of the strains they hold, and identification services for unknown organisms. Frequently, the catalogues will be computerized, so that screens for strains with specific activities can be readily achieved. In the UK, the development of on-line computer access to data from all the culture collections is being considered.

The recent and expanding interest in biotechnology is heavily dependent on culture collections, firstly as a source of useful organisms and also for the deposition of cultures. If a process involving a microorganism is patented, a culture of that strain must be deposited in a culture collection that is an accepted International Depository Authority. Moreover, culture collections offer other services to industry, including the preservation of production strains, strain selection programmes and, as mentioned previously, identification of unknown organisms. Culture collections are indispensable to microbiology, and it is encouraging that international cooperation is leading to the maintenance of a wider range of microorganisms and cell lines in collections. Moreover computerized data handling and on-line access to the mass of information housed in these collections will make them increasingly useful (see Kirsop, 1983).

REFERENCES

Brisbane, P. G. and Rovira, A. D. (1961). A comparison of methods for classifying rhizosphere bacteria, *Journal of General Microbiology*, **26**, 379 – 392.

Cowan, S. T. (1974). *Cowan and Steel's Manual for the Identification of Medical Bacteria*, 2nd edn. Cambridge University Press, Cambridge.

Glover, D. M. (1984). *Gene Cloning, the Mechanics of DNA Manipulation*. Chapman and Hall, London.

Gray, T. R. G. (1969). The identification of soil bacteria, in *The Soil Ecosystem* (J. G. Sheals, Ed.) pp. 73 – 82. The Systematics Association Publication No. 8.

Hissett, R. and Gray, T. R. G. (1974). Bacterial populations of litter and soil in a decidious woodland. I. Qualitative studies, *Revue d'ecologie et de Biologie du Sol*, **10**, 495 – 508.

Kirsop, B. (1983). Culture collections-their services to biotechnology, *Trends in Biotechnology*, **1**, 4 – 8.

Last, F. T. and Deighton, F. C. (1965). The non-parasitic microflora on the surface of living leaves, *Transactions of the British Mycological Society* **48**, 83 – 99.

Lawrence, D. R. and Priest, F. G. (1981). Identification of brewery cocci, in *Proceedings of the European Brewery Convention, Copenhagen*, pp. 217 – 227.

Lowe, W. E. and Gray, T. R. G. (1972). Ecological studies on coccoid bacteria in a pine forest soil. I. Classification, *Soil Biology and Biochemistry*, **4**, 459 – 467.

Lowe, W. E. and Gray, T. R. G. (1973a). Ecological studies on coccoid bacteria in a pine forest soil II. Growth of bacteria inoculated into soil, *Soil Biology and Biochemistry*, **5**, 449 – 452.

Lowe, W. E. and Gray, T. R. G. (1973b). Ecological studies on coccoid bacteria in a pine forest soil. III. Competition interactions between bacterial strains in soil, *Soil Biology and Biochemistry*, **5**, 463 – 472.

O'Brien, S. J. (1984). *Genetic Maps: a Compilation of Linkage and Restriction Maps of Genetically Studied Organisms*, Vol 3. Cold Spring Harbor Laboratory, New York.

Ruinen, J. (1961). The phyllosphere. I. An ecologically neglected milieu, *Plant and Soil*, **15**, 81 – 109.

Williams, S. T., Goodfellow, M. and Vickers, J. C. (1984). New microbes from old habitats?, *Symposium of the Society for General Microbiology*, **36**(II), 219 – 256.

8 Conclusions and outlook

Since its origins as a science in the nineteenth century, bacterial taxonomy has enjoyed a phase of rising interest with a concomitant increase in knowledge, particularly in the years since the end of the Second World War. New techniques, such as chemotaxonomy and numerical taxonomy, have been developed, their value being attested by their widespread use throughout the scientific community. Thus within only a few decades bacterial taxonomy has progressed from a comparatively mystical craft to a respectable exacting science.

The question is therefore appropriate about what developments seem likely to follow in the future. Of course predicting the future is a notoriously unreliable business, but within the realms of bacterial taxonomy there are some trends which should reach fruition within the next decade. For example, it is to be hoped that there will be a greater degree of standardization in methodology.

Within the subject of numerical taxonomy, there should be agreement on the nature of the testing regimes and about the subsequent computer analyses. Questions to be resolved include the number and nature of the tests to be used; at present numerical taxonomy studies employ a wide range of tests, numbering between 50 and 200. Moreover, there is very little consistency between studies; therefore, direct comparisons between the multitude of publications is extremely difficult. For example, some studies may highlight morphological features whereas others are biased towards carbon utilization tests. In addition, there are numerous methods and media for ascertaining the ability of organisms to utilize compounds as the sole source of carbon for energy and growth. Agreement needs to be reached about the precise methods to be used, although in many cases this is a problematical task. Thus, bacteria differ widely in their physiology, and tests for sugar fermentations or urease activity developed for the Enterobacteriaceae do not necessarily give meaningful information if applied to other groups, such as *Pseudomonas* or *Bacillus*. So the requirement for standardization of phenotypic testing must be consistent with the physiology of the organisms in question. Nevertheless, it may be expected that some of the less reliable, poorly defined conventional tests, e.g. the Voges Proskauer reaction, will be replaced by better methods. Fortunately, increasingly sensitive techniques are being devised for the characterization of sub-cellular components, many of which have taxonomic value. Clearly, the message is that advances in taxonomy depend upon developments in allied disciplines.

Standardization might also be reached on the nature of the computer analyses to be carried out in numerical taxonomy studies. For example, it could be argued that certain coefficients and clustering techniques should be adopted for all investigations. Already, the majority of publications involve use of the Jaccard and/ or simple matching coefficient in combination with average- or single-linkage clustering, but again variation among taxa must be accommodated. For example, in some recent numerical analyses of lactobacilli, the high phenetic similarity among strains resulted in poor clustering when using unweighted average linkage clustering. Therefore, it was necessary to resort to other algorithms, notably complete linkage clustering to generate discrete clusters (see Priest and Barbour, 1985). It is a pity that interpretation of the analyses seems to be such a haphazard affair, with taxa (these are usually equated with species) defined at an arbitrary point between the 70 and 90% similarity levels. Thus a 'species' in one study could well equate with a 'tribe' in another publication; albeit with a different group of organisms. This situation will probably only be resolved by reference to chemotaxonomic information.

It is anticipated that the current interest in chemotaxonomy and molecular genetics will continue. However, care needs to be given to the degree of emphasis placed upon such information in the formulation of classifications. At present, many Gram-positive taxa are defined largely on the basis of chemotaxonomic characters, and it remains to be seen whether or not this is sound policy. It would appear to be more beneficial for many such characters to be incorporated in numerical taxonomy studies; otherwise, undue weight will be given to what is, after all, often limited information on the biology of the organisms under study. Classifications based largely on DNA : DNA hybridizations escape from this criticism because the complete genetic information of the bacteria in question is being used. Moreover, as stressed in Chapter 3, the composition of nucleic acids is totally unaffected by the environment, and any organism may be compared with any other under standard conditions. Thus nucleic acid analyses are perhaps the only source of standardized data for the delineation of taxa, although arguments concerning the level of genetic heterogeneity allowable within a species still flourish. On similar grounds, rRNA analyses seem to be the most appropriate criteria for the delineation of higher ranked taxa, such as genera and families. It is important that such classifications are consistent with those based on the phenotype and it is heartening that, in general, this is the case.

With the publication of the Approved Lists of Bacterial Names in 1980, bacterial nomenclature became a reasonably orderly process and, at last, identification is being treated with the importance it deserves. The conventional approaches of morphological and biochemical tests and the paraphernalia of dichotomous keys and/or diagnostic tables are gradually being replaced by commercial kits and computerized data bases for clinically important bacteria; but for most non-medical organisms only the traditional schemes are available. Fortunately, the availability of computer software is enabling those scientists interested in disciplines other than medical microbiology to generate probabilistic identification matrices for various bacterial groups. Such matrices will be included in future editions of *Bergey's Manual of Systematic Bacteriology* for general usage.

Although we forsee identification of bacteria largely relying on phenotypic testing in the near future, other methods are being developed and appear to be gradually gaining acceptance. For example, hybridization probes, based on [32]P- labelled DNA, are being used to detect toxigenic *E. coli* strains in faecal samples. Speed and sensitivity are two advantages of this system, and with the introduction of non-radioactive detection of the hybrids, it will become more readily acceptable. Other chemotaxonomic criteria that may be used for identification purposes include lipids, whole cell protein patterns and the pyrolysis products of complete organisms.

Serology has been revolutionized by the advent of highly specific monoclonal antibodies and their incorporation into sensitive serological techniques, such as the enzyme linked immunosorbent assay (ELISA). Reliable rapid diagnoses are possible with these systems, and there is not a need for expensive equipment. Indeed, they can often be used in the field to examine diseased animals or plants for the presence of specific organisms. Techniques such as ELISA are likely to be the subject of great interest by commerce in the quest for diagnostic kits for the mass markets. Not only will this interest improve diagnosis of disease but the techniques should also aid industry in recognizing particular strains. For example, these could have industrial significance in the production of valuable compounds, such as antibiotics and enzymes. The ability to recognize such strains accurately would quickly identify rogue organizations which might pirate the cultures for competitive reasons. The policing of such blatant industrial espionage could be a valuable offshoot of taxonomic development.

In short, the message is that bacterial taxonomy is not a subject doomed to museums. Instead, taxonomy is an exciting branch of science with considerable promise for the future. We await, with considerable interest, the next important landmark in the development of bacterial taxonomy. Such activities of tomorrow will result from the students of today. Therefore, the kindling of interest among our readers may generate the ideas to replace the currently popular numerical taxonomy and chemotaxonomy methods.

REFERENCE

Priest, F. G. and Barbour, E. A. (1985). Numerical taxonomy of lactic acid bacteria and some related taxa, in *Computer-assisted Bacterial Systematics*, (M. Goodfellow, D. Jones and F. G. Priest, Eds.), pp. 137 – 164, Academic Press, London.

Glossary

Additive coding   A system of coding quantitative features into binary code, which retains the scale of the character.

Algorithm   A set of instructions for a series of calculations.

A posteriori (**L.**)   'From what comes after', i.e. inductively.

Approved Lists of Bacterial Names   A list of validly published bacterial names that appeared in the January 1980 issue of the *International Journal of Systematic Bacteriology*. Any names not included in these or subsequent updates are not nomenclaturally valid.

A priori (**L.**)   'From what is before', i.e. presumptively.

Binary character   A character existing in two states, as plus (+) or negative (−).

Binomial name   Owing its origin to the Swedish naturalist Linnaeus, the name of a species comprises two words, with the first being the genus name, and the second being the species name, i.e. specific epithet.

Canonical variates analysis   See Multiple discrimination analysis.

Centroid   A point representing the centre of a cluster.

Centrotype   The nearest OTU to the centroid.

Character   From the taxonomic standpoint, the term refers to any measurable property.

Chemotaxonomy   Taxonomy based upon chemical analyses of whole cells or subcellular components.

Chemotype   A representative strain for a set of chemical characteristics.

Cladistic   Refers to the branching pattern that describes the pathway of ancestry of a group of organisms.

Cladogram   A tree-like diagram that represents the evolutionary pathway of a group of organisms.

Classification   The process of ordering organisms into groups.

Cluster analysis   The mathematical determination of groups (clusters) from a distance or similarity matrix.

Coefficient   A measure used to calculate the similarity or distance between two OTUs.

Combinatio nova (*comb. nov.*)   New combination, applies to a name which

119

may result from, for example, transfer of a species from one genus to another resulting from improvements in its taxonomy.

Congruence   The level of agreement between two taxonomic methods.

Convergence   The coming together of two distinct evolutionary paths.

Cophenetic correlation   Measure of the accuracy with which a dendrogram represents a similarity matrix.

Dendrogram   A tree-like diagram resulting from cluster analyses, which expresses relationships between OTUs.

Diagnosis   A brief description of a taxon, permitting its differentiation from all others. The word is often used in the context of disease.

Dichotomous (diagnostic) keys   Identification keys in which the sequential answering of questions ultimately provides an identification.

Discriminant analysis   *See* Multiple discrimination analysis.

Dissimilarity   The exact opposite of similarity.

Divergence index   An indication of sequence divergence established by dividing DNA reassociation measured at a stringent temperature by that at the optimal temperature.

Equal weight   Appertains to the notion that in numerical classification, all characters (tests) are of the same importance.

Euclidean distance   The measurement of distance between OTUs in terms of their coordinates on right-angled (Cartesian) axes.

Genus novum (gen. nov.)   New genus.

Hierarchy   An organization with ranks, graded one above the other.

Holotype   A type strain reported by the original author.

Hypothetical median organism   These are so-called 'average' organisms and are representative of clusters of OTUs. The term was coined by Liston and co-workers in 1963.

Hypothetical taxonomic units   OTUs with theoretical characteristics used in the construction of Wagner trees (cladograms).

Identification   The process by which organisms are assigned to taxonomic groups, which have been previously defined as a result of classification.

Illigitimate name   A name not published in accordance with the International Code of Nomenclature of Bacteria.

International Code of Nomenclature of Bacteria   A formal code, originally mooted at the Second International Congress of Microbiology in 1936, which seeks to ensure the presence of stable, meaningful bacterial names.

Incongruence   The opposite of congruence, i.e. the level of disagreement between two taxonomies.

Isochronic   In the present context, the term refers to individuals within a group evolving at a similar rate.

Judicial Commission   Approved by the Second International Congress for Microbiology in 1936 to issue formal nomenclatural 'Opinions', upon request.

Lectotype   If a subsequent author designates one of the original author's strains as a type culture, it is referred to as the lectotype.

Legitimate name   A name published in accordance with the International Code of Nomenclature of Bacteria.

Matrix (similarity or dissimilarity)   In numerical taxonomy, this is a triangular array of similarity or distance estimates, which permits the comparison of an OTU with any other listed in the data set.

Metric   An analogous term for dissimilarity measurements.

Minimal evolution   The concept that evolution proceeds along the shortest possible pathway with the fewest number of steps (see parsimony).

Monophyletic group   A phylogenetic group based on the shared possession of a homologous character.

Monothetic   A group, membership of which is dependent on the presence or absence of a few invariant characters.

Monotype   Where an original author described only one strain, which was not specifically regarded as the holotype, it is regarded as the monotype.

Multistate characters   Characters/tests which occur in more than one state, e.g. tests which may be scored as negative, weakly-, moderately-, or strongly-positive and could be recorded numerically as '0', '1', '2' or '3', respectively.

Multiple discrimination analysis   Similar to Principal component analysis, but the axes represent the greatest variability of the means of the different taxa. Used for identification.

Natural (phenetic) classification   Natural, in this connotation, refers to classifications based upon as many aspects of the biology of the organisms as possible.

Neotype   A culture of the original author's batch designated as a type culture by a subsequent author.

Nomenclature   The process of allocating names to groups of organisms.

Nomenclatural type   *See* Type strain.

Nomen conservandum   Conserved name—as voiced through an Opinion of the Judicial Commission. Conserved names *must* be used instead of synonyms.

Nomen rejiciendum   Rejected name—as voiced through an Opinion of the Judicidial Commission.

Numerical taxonomy   The arrangement of organisms into groups on the basis of their overall characteristics, as a result of numerical methods.

Operational taxonomic unit (OTU)   Within the realms of bacteriology, refers to the bacterial isolate/strain which is the subject of study.

'Opinion'   A formal view on a matter of nomenclature expressed by the Judicial Commission.

Ordination methods   Mathematical methods which reduce the number of dimensions in a taxonomic space to two or three, such that distance between OTUs can be represented.

Parsimony   The theory that evolution proceeds via the fewest number of steps.

Pattern difference   As discussed by Professor P.H.A. Sneath, the term applies to one of the two components of the total difference between OTUs. Pattern difference (D_p) may be expressed as: $D_p = [2\sqrt{(bc)}]/(a + b + c + d)$ (see Chapter 2 for interpretation of symbols).

Phenetic classification   A classification based on the overall properties of the organisms, as they are percieved at present.

Phenon (pl. Phena)   A group defined on the basis of high overall similarity of all of its members.

Phenotype   Appertains to observable properties, without regard to ancestry.

Phylogeny   The study of the evolution/ancestry of organisms.

Polythetic   Refers to a group, characterized by a high number of common features (characters). Membership of the group does not require the presence or absence of any particular attribute. In this context, phena are polythetic.

Principal components analysis   An ordination method in which the first principal component is derived as a dimension representing the greatest variability among the data and plotted against a second dimension which accounts for the next greatest variability.

Principal coordinate analysis   An ordination method similar to Principal components analysis but applicable to taxonomic data not necessarily based on Euclidean distance. If the distances are Euclidean, it becomes the same as Principal components analysis.

Priority   Within the terms of the Bacteriological Code of Nomenclature, priority is given to the first valid publication of a name.

Scatter diagram   A representation of the position of OTUs in a two-dimensional taxonomic space.

Similarity   In the context of numerical taxonomy, the term refers to the level of resemblance between OTUs.

Species   The basic taxonomic unit, which *should* consist of highly related isolates.

Species nova (sp. nov.)   New species.

Specific epithet   The second word of a binomial name, indicating the identity of a species.

Systematics   The study of the diversity and relationship among organisms.

Taxometrics   *See* Numerical taxonomy.

Taxon (pl. taxa)   A group of individuals of any rank.

Taxonomic map   The two- or three-dimensional arrays showing the ordering of OTUs as a result of ordination methods.

Taxonomy   The theory of classification, nomenclature and identification.

Taxospecies   A group of isolates with a mutually high phenetic similarity.

Test error/reproducibility   A term first coined by Professor P.H.A. Sneath for the mistakes which may creep into the recording of characters. Unfortunately, the definition of positive and negative responses of some tests is vague.

Thermal binding index   *See* Divergence index.

Transformed cladistics   Classifications based on maximum parsimony but with no recourse to evolutionary theory.

Type (reference) strain   An organism/culture which has been designated as a permanent representative of a taxon. Type strains will be deposited in culture collections.

Ultrametric property   The fact that a cladogram will be the same as a phenogram/dendrogram if evolutionary rates are constant and convergence minimal.

Unit character   A taxonomic character (test) of two or more states, which cannot be subdivided logically except for changes in the method of coding.

Vigour difference   As discussed by Professor P.H.A. Sneath, this term applies to the differences in growth rates between OTUs. Mathematically, the vigour difference (D_v) may be expressed as: $D_v = (c - b)/(a + b + c + d)$ (see Chapter 2 for the meaning of the symbols).

Wagner trees   Minimal length trees used to represent most parsimonious cladograms.

Appendix 1
Classification of bacteria

This scheme has been based on data from Bergey's *Manual of Determinative Bacteriology*, 8th edn. (1974), Bergey's *Manual of Systematic Bacteriology*, Vol. 1 (1984), and the *Approved Lists of Bacterial Names* (1980) and their supplements.

⋆ refers to the type genus (see Chapter 5). The type species, for each genus, is given in parentheses.

It should be emphasized that, unlike botanical and zoological classifications, bacteria do not all fit into a convenient hierarchical arrangement, culminating in a single hyper-group at the tip of a pyramid. This point will become abundantly clear. Nevertheless for convenience, the bacteria are considered to be divided as follows:

Kingdom: Procaryotae
Division: Gracilicutes (Gram-negative bacteria)
  Class: Scotobacteria
  Class: Anoxyphotobacteria
  Class: Oxyphotobacteria
Division: Firmicutes (Gram-positive bacteria)
  Class: Firmibacteria
  Class: Thallobacteria
Division: Tenericutes (Bacteria, which lack rigid cell walls)
  Class: Mollicutes
Division: Mendosicutes (Bacteria with unusual cell walls)
  Class: Archaeobacteria

Within this framework all bacteria should be placed. However to simplify matters, the bacteria have been grouped according to certain traits:

Phototrophic Bacteria

Order: Rhodospirillales
Family: Chromatiaceae
Genus: *Chromatium* (*C. okenii*),   *Amoebobacter* (*A. roseus*),
                   Lamprobacter (*L. modestohalophilus*)   *Lamprocystis* (*L. roseopersicina*)
                   Thiocystis (*T. violacea*)   *Thiodictyon* (*T. elegans*)
                   Thiopedia (*T. rosea*)   *Thiospirillum* (*T. jenense*)

Family: Chlorobiaceae
Genus: *Chlorobium* (*C. limicola*)   *Chloroherpeton* (*C. thalassium*)
                   Pelodictyon (*P. clathratiforme*) *Prosthecochloris* (*P. aestuarii*)
Related genus: *Heliobacterium* (*H. chlorum*)

Family: Ectothiorhodospiraceae
Genus: *Ectothiorhodospira* (*E. mobilis*)

Family: Rhodospirillaceae
Genus: *Rhodospirillum* (*R. rubrum*)   *Rhodobacter* (*R. capsulatus*)
                   Rhodomicrobium (*R. vannielii*)   *Rhodocyclus* (*R. purpureus*)
                   Rhodopseudomonas (*R. palustris*) *Rhodopila* (*R. globiformis*)
Related genus: *Erythrobacter* (*E. longus*)

Gliding bacteria

Order: Myxobacterales
Family: Archangiaceae
Genus: *Archangium (A. gephyra)
Family: Cystobacteraceae
Genus: *Cystobacter (C. fuscus)
Stigmatella (S. aurantiaca)
Melittangium (M. boletus)
Family: Polyangiaceae
Genus: *Polyangium (P. vitellinum)
Nannocystis (N. exedens)
Chondromyces (C. crocatus)

Order: Cytophagales
Family: Cytophagaceae
Genus: *Cytophaga (C. hutchinsonii)
Flexithrix (F. dorotheae)
Saprospira (S. grandis)
Flexibacter (F. flexilis)
Herpetosiphon (H. aurantiacus)
Sporocytophaga (S. myxococcoides)
Chitinophaga (C. pinensis)
Sphingobacterium (S. spiritivorum)
Related genus: Capnocytophaga (C. ochracea)
Microscilla (M. marina)
Family: Leucotrichaceae
Genus: *Leucothrix (L. mucor)
Thiothrix (T. nivea)
Family: Simonsiellaceae
Genus: *Simonsiella (S. muelleri)
Alysiella (A. filiformis)

Order: Lysobacterales
Family: Lysobacteraceae
Genus: *Lysobacter (L. enzymogenes)

Order: Myxococcales
Family: Myxococcaceae
Genus: *Myxococcus (M. fulvus)
Angiococcus (A. disciformis)

Families of uncertain affiliation:
Family: Achromatiaceae
  Genus: *Achromatium (A. oxaliferum)
Family: Vitreoscillaceae
  Genus: *Vitreoscilla (V. beggiatoides)
Genera of uncertain affiliation:
             Filibacter (F. limicolor)   Toxothrix (T. trichogenes)

Sheathed bacteria
Family: Crenothrichaceae
  Genus: *Crenotrix (C. polyspora)
Genera of uncertain affiliation:
             Blastococcus (B. aggregatus)   Haliscomenobacter (H. hydrossis)
             Leptothrix (L. ochracea)   Sphaerotilus (S. natans)

Budding and appendaged bacteria

Order: Caulobacterales
Family: Caulobacteraceae
  Genus: *Caulobacter (C. vibrioides)
  Related genus: Prosthecobacter (P. fusiformis)

Order: Hyphomicrobiales
Family: Hyphomicrobiaceae
  Genus: *Hyphomicrobium (H. vulgare)   Hyphomonas (H. polymorpha)
  Related genus: Pedomicrobium (P. ferrugineum)
Families of uncertain affiliation:
Family: Gallionellaceae
  Genus: *Gallionella (G. ferruginea)
Family: Nevskiaceae
  Genus: *Nevskia (N. ramosa)

Family: Pasteuriaceae
Genus: *Pasteuria (P. ramosa)
Related genus: Pirella (P. staleyi)
Genera of uncertain affiliation:
    Ancalomicrobium (A. adetum)   Asticcacaulis (A. excentricus)
    Blastobacter (B. henricii)   Ensifer (E. adhaerens)
    Gemmata (G. obscuriglobus)   Labrys (L. monahos)
    Planctomyces (P. bekefii)   Prosthecomicrobium (P. pneumaticum)
    Seliberia (S. stellata)

Aerobic/micro-aerophilic non-motile/motile helical/vibrioid Gram-negative bacteria

Order: Spirillales
Family: Spirillaceae
Genus: *Spirillum (S. volutans)   Campylobacter (C. fetus)
    Aquaspirillum (A. serpens)   Azospirillum (A. lipoferum)
    Bdellovibrio (B. bacteriovorus)   Oceanospirillum (O. linum)
    Vampirovibrio (V. chlorellavorus)

Family of uncertain affiliation:
Family: Spirosomaceae
Genus: *Spirosoma (S. linguale)   Flectobacillus (F. major)
    Runella (R. slithyformis)
Related genus: Ancylobacter (A. aquaticus)   Brachyarcus (B. thiophilus)
    Meniscus (M. glaucopsis)   Pelosigma (P. cohnii)

Gram-negative aerobic rods and cocci

Family:	Acetobacteriaceae	
Genus:	*Acetobacter (A. aceti)*	Gluconobacter (G. oxydans)
Family:	Azotobacteriaceae	
Genus:	*Azotobacter (A. chroococcum)*	Azomonas (A. agilis)
	Azomonotrichon (A. macrocytogenes)	Azorhizophilus (A. paspali)
	Beijerinckia (B. indica)	Derxia (D. gummosa)
Family:	Brucellaceae	
Genus:	*Brucella (B. melitensis)*	
Family:	Halobacteriaceae	
Genus:	*Halobacterium (H. salinarium)*	Halococcus (H. morrhuae)
Family:	Legionellaceae	
Genus:	*Legionella (L. pneumophila)*	Fluoribacter (F. bozemanae)
	Tatlockia (T. micdadei)	
Family:	Neisseriaceae	
Genus:	*Neisseria (N. gonorrhoeae)*,	Acinetobacter (A. calcoaceticus)
	Branhamella (B. catarrhalis)	Kingella (K. kingae)
	Moraxella (M. lacunata)	Morococcus (M. cerebrosus)
Family:	Rhizobiaceae	
Genus:	*Rhizobium (R. leguminosarum)*	Agrobacterium (A. tumefaciens)
	Bradyrhizobium (B. japonicum)	Phyllobacterium (P. myrsinacearum)
Family:	Methylococcaceae	
Genus:	*Methylococcus (M. capsulatus)*	Methylomonas (M. methanica)
Related Genus:	Methylobacillus (M. glycogenes)	Methylobacterium (M. organophilum)
	Methylophaga (M. marina)	Protomonas (P. extorquens)
Genus of uncertain affiliation:		
	Agitococcus (A. lubricus)	

Order: Pseudomonadales
Family: Pseudomonadaceae
Genus: *Pseudomonas (P. aeruginosa)*
Xanthomonas (X. campestris)

Genera of uncertain affiliation:

Achromobacter (A. xylosoxidans)
Agromonas (A. oligotrophica)
Alteromonas (A. macleodii)
Deleya (D. aesta)
Francisella (F. tularensis)
Lampropedia (L. hyalina)
Paracoccus (P. denitrificans)
Serpens (S. flexibilis)
Thermus (T. aquaticus)
Natronobacterium (N. gregoryi)

Acidiphilium (A. cryptum)
Alcaligenes (A. faecalis)
Bordetella (B. pertussis)
Flavobacterium (F. aquatile)
Janthinobacterium (J. lividum)
Marinomonas (M. communis)
Phenylobacterium (P. immobile)
Thermomicrobium (T. roseum)
Xanthobacter (X. autotrophicus)
Natronococcus (N. occultus)

Frateuria (F. aurantia)
Zoogloea (Z. ramigera)

Facultatively anaerobic Gram-negative rods
Family: Enterobacteriaceae
Genus: *Escherichia (E. coli)*
Buttiauxella (B. agrestis)
Citrobacter (C. freundii)
Enterobacter (E. cloacae)
Erwingella (E. americana)
Klebsiella (K. pneumoniae)
Koserella (K. trabulsii)
Levinea (L. amalonatica)
Morganella (M. morganii)
Pectobacterium (P. carotovorum)
Providencia (P. alcalifaciens)
Salmonella (S. choleraesuis)

Budvicia (B. aquatica)
Cedecea (C. davisae)
Edwardsiella (E. tarda)
Erwinia (E. amylovora)
Hafnia (H. alvei)
Kluyvera (K. ascorbata)
Leminorella (L. grimontii)
Moellerella (M. wisconsensis)
Obesumbacterium (O. proteus)
Proteus (P. vulgaris)
Rahnella (R. aquatilis)
Serratia (S. marcescens)

Shigella (S. dysenteriae)
Xenorhabdus (X. nematophilus)
Yokenella (Y. regensburgei)
Tatumella (T. ptyseos)
Yersinia (Y. pestis)

Family:
Genus: Pasteurellaceae
*Pasteurella (P. multocida)
Haemophilus (H. influenzae)
Actinobacillus (A. lignieresii)

Related genus: Gardnerella (G. vaginalis)
Taylorella (T. equigenitalis)

Family: Vibrionaceae
Genus: *Vibrio (V. cholerae)
Allomonas (A. enterica)
Photobacterium (P. phosphoreum)
Aeromonas (A. hydrophila)
Halomonas (H. elongata)
Plesiomonas (P. shigelloides)

Genera of uncertain affiliations:
Calymmatobacterium (C. granulomatis)
Chromobacterium (C. violaceum)
Streptobacillus (S. moniliformis)
Cardiobacterium (C. hominis)
Eikenella (E. corrodens)
Zymomonas (Z. mobilis)

Gram-negative anaerobic rods and cocci

Family: Bacteroidaceae
Genus: *Bacteroides (B. fragilis)
Acetoanaerobium (A. noterae)
Acidaminobacter (A. hydrogenoformans)
Anaerovibrio (A. lipolytica)
Centipeda (C. periodontii)
Lachnospira (L. multiparis)
Pectinatus (P. cerevisiiphilus)
Propionigenium (P. modestum)
Rikenella (R. microfusus)
Selenomonas (S. sputigena)
Succinivibrio (S. dextrinosolvens)
Syntrophus (S. buswellii)
Acetivibrio (A. cellulolyticus)
Acetomicrobium (A. flavidum)
Anaerobiospirillum (A. succiniproducens)
Butyrivibrio (B. fibrisolvens)
Fusobacterium (F. nucleatum)
Leptotrichia (L. buccalis)
Pelobacter (P. acidigallici)
Propionispira (P. arboris)
Roseburia (R. cecicola)
Succinimonas (S. amylolytica)
Syntrophomonas (S. wolfei)
Wolinella (W. succinogenes)

Related genus: *Capsularis (C. zoogleiformans)*
Mitsuokella (M. multiacidus)
Megamonas (M. herpermegas)
Mobiluncus (M. curtisii)

Family: Veillonellaceae
Genus: *Veillonella (V. parvula)*
Megasphaera (M. elsdenii)
Acidaminococcus (A. fermentans)

Related genus: *Gemmiger (G. formicilis)*
Family: Haloanaerobiaceae
Genus: *Haloanaerobium (H. praevalens)*
Halobacteroides (H. halobius)

Family: Desulfurococcaceae
Genus: *Delsulfurococcus (D. mucosus)*
Related genus: *Desulfobacter (D. postgatei)*
Desulfococcus (D. multivorans)
Desulfonema (D. limicola)
Desulfovibrio (D. desulfuricans)
Desulfobulbus (D. propionicus)
Desulfomonas (D. pigra)
Desulfosarina (D. variabilis)
Desulfuromonas (D. acetoxidans)

Genera of uncertain affiliation:
Syntrophobacter (S. wolinii)
Thermobacteroides (T. acetoethylicus)

Miscellaneous anaerobes
Genus: *Llyobacter (L. polytropus)*
Oxalobacter (O. formigenes)

Miscellaneous thermophilic Gram-negative bacteria

Order: Thermoproteales
Family: Thermoproteaceae
Genus: *Thermoproteus (T. tenax)*
Thermofilum (T. pendens)
Thermococcus (T. celer)

Gram-negative chemolithotrophs

Ammonia or nitrite oxidizing bacteria
Family: Nitrobacteraceae
Genus: *Nitrobacter (N. winogradskyi)*
             Nitrococcus (N. nitrosus)
             Nitrosomonas (N. europaea)
             Nitrospina (N. gracilis)
             Nitrococcus (N. mobilis)
             Nitrosolobus (N. multiformis)
             Nitrosospira (N. briensis)

Sulphur bacteria
Genera of uncertain affiliation:
             Macromonas (M. mobilis)
             Thermothrix (T. thiopara)
             Thiomicrospira (T. pelophila)
             Thiospira (T. winogradskyi)
             Pyrodictium (P. occultum)
             Sulfolobus (S. acidocaldarius)
             Thiobacillus (T. thioparus)
             Thiosphaera (T. pantotropha)
             Thiovulum (T. majus)

Family: Thiocapsaceae
Genus: *Thiocapsa (T. roseopersicina)*

Hydrogen bacteria

Genus: *Hydrogenobacter (H. thermophilus)*
             Calderobacterium (C. hydrogenophilum)

Methane bacteria

Order: Methanobacteriales
Family: Methanobacteriaceae
Genus: *Methanobacterium (M. formicicum)*
Related genus: *Methanobrevibacter (M. ruminatium)*
             Methanogenium (M. cariaci)
             Methanospirillum (M. hungatii)

Order: Methanococcales
Family: Methanococcaceae
Genus: *Methanococcus (M. mazei)*

Order: Methanomicrobiales
Family: Methanomicrobiaceae
Genus: *Methanomicrobium (M. mobile)*
Methanothrix (M. soehngenii)

Families of uncertain affiliation:
Family: Methanosarcinaceae
Genus: *Methanosarcina (M. methanica)*   *Methanococcoides (M. methyluteus)*
Family: Methanoplanaceae   *Methanolobus (M. tindarius)*
Genus: *Methanoplanus (M. limicola)*
Family: Methanothermaceae
Genus: *Methanothermus (M. fervidus)*

Gram-positive cocci

Aerobic and/or facultatively anaerobic cocci

Order: Micrococcales
Family: Micrococcaceae
Genus: *Micrococcus (M. luteus)*
Staphylococcus (S. aureus)
Paracoccus (P. denitrificans)
Related genus: Planococcaceae   *Stomatococcus (S. mucilaginosus)*
Family: Planococcaceae
Genus: *Planococcus (P. citreus)*

Family: Deinococcaceae
Genus: *Deinococcus (D. radiodurans)
Family of uncertain affiliation:
Family: Streptococcaceae
Genus: *Streptococcus (S. pyogenes)
             Enterococcus (E. faecalis)
             Leuconostoc (L. mesenteroides)
             Trichococcus (T. flocculiformis)
             Aerococcus (A. viridans)
             Gemella (G. haemolysans)
             Pediococcus (P. damnosus)
Related genus: Melisococcus (M. pluton)
Genus of uncertain affiliation:
             Saccharococcus (S. thermophilus)

Anaerobic cocci
Family: Peptococcaceae
Genus: *Peptococcus (P. niger)
             Peptostreptococcus (P. anaerobius)
             Sarcina (S. ventriculi)
             Coprococcus (C. eutactus)
             Ruminococcus (R. flavefaciens)

Endospore-forming rods and cocci

Order: Bacillales
Family: Bacillaceae
Genus: *Bacillus (B. subtilis)
              Sporolactobacillus (S. inulinus)
              Desulfotomaculum (D. nigrificans)
              Sporosarcina (S. ureae)

Order: Clostridiales
Family: Clostridiaceae
Genus: *Clostridium (C. butyricum)
Family of uncertain affiliation:
Family: Oscillospiraceae
Genus: *Oscillospira (O. guilliermondi)

Gram-positive non-sporing rods

Family: Lactobacillaceae
Genus: *Lactobacillus (L. delbreuckii)
Brochothrix (B. thermosphacta)
Listeria (L. monocytogenes)
Related genus: Erysipelothrix (E. rhusiopathiae)
Genera of uncertain affiliation: Falcivibrio (F. grandis)
Thermoanaerobium (T. brockii)
Thermoanaerobacter (T. ethanolicus)

Actinomycetes and related bacteria

Order: Actinomycetales
Family: Actinomycetaceae
Genus: *Actinomyces (A. bovis)
Bacterionema (B. matruchoti)
Rothia (R. dentocariosa)
Arachnia (A. propionica)
Bifidobacterium (B. bifidum)

Family: Actinoplanaceae
Genus: *Actinoplanes (A. philippinensis)
Ampullariella (A. regularis)
Kitasatoa (K. purpurea)
Planobispora (P. longispora)
Spirillospora (S. albida)
Amorphosporangium (A. auranticolor)
Dactylosporangium (D. aurantiacum)
Pilimelia (P. terevasa)
Planomonospora (P. parontospora)
Streptosporangium (S. roseum)

Family: Dermatophilaceae
Genus: *Dermatophilus (D. congolensis)
Geodermatophilus (G. obscurus)
Related genus: Conglomeromonas (C. largomobilis)

Family: Frankiaceae
Genus: *Frankia (F. alni)

Family: Micromonosporaceae
Genus: *Micromonospora (M. chalcea)
Micropolyspora (M. brevicatena)
Thermomonospora (T. curvata)
Microbispora (M. rosea)
Thermoactinomyces (T. vulgaris)

Related genus: *Faenia (F. rectivirgula)*   *Promicromonospora (P. citrea)*

Family: Nocardiaceae
Genus: **Nocardia (N. asteroides)*   *Actinopolyspora (A. halophila)*
     Pseudonocardia (P. thermophila)   *Saccharopolyspora (S. hirsuta)*

Family: Streptomycetaceae
Genus: **Streptomyces (S. albus)*   *Actinopycnidium (A. caeruleum)*
     Actinosporangium (A. violaceum)   *Microellobosporia (M. cinerea)*
     Nocardioides (N. albus)   *Sporochthya (S. polymorpha)*
     Streptoverticillium (S. baldaccii)

Related genus: *Chainia (C. antibiotica)*   *Elytrosporangium (E. brasiliensis)*
     Intrasporangium (I. calvum)

Order: Mycobacteriales
Family: Mycobacteriaceae
Genus: **Mycobacterium (M. tuberculosis)*

Order: Caryophanales
Family: Caryophanaceae
Genus: **Caryophanon (C. latum)*

Genera of uncertain affiliation:
     Actinomadura (A. madurae)   *Actinosynnema (A. mirum)*
     Kineosporia (K. aurantiaca)   *Kitasatosporia (K. setalba)*
     Microtetraspora (M. glauca)   *Nocardiopsis (N. dassonvillei)*
     Renibacterium (R. salmoninarum)   *Saccharomonospora (S. viridis)*
     Saccharothrix (S. australiensis)

Coryneform bacteria
Family: Brevibacteriaceae
Genus: **Brevibacterium (B. linens)*
Family: Corynebacteriaceae
Genus: **Corynebacterium (C. diphtheriae)*
Family: Propionibacteriaceae
Genus: **Propionibacterium (P. freundenreichii)*   *Acetobacterium (A. woodii)*

Order:
Tribe: Eubacteriales
            Eubacterieae
Genus: *Eubacterium (E. foedans)
Genera of uncertain affiliation:

Arcanobacterium (A. haemolyticum)
Aureobacterium (A. liquefaciens)
Cellulomonas (C. flavigena)
Curtobacterium (C. citreum)
Kurthia (K. zopfii)
Pimelobacter (P. simplex)

Arthrobacter (A. globiformis)
Caseobacter (C. polymorphus)
Clavibacter (C. michiganense)
Exiguobacterium (E. aurantiacum)
Microbacterium (M. lacticum)

The spirochaetes

Order: Spirochaetales
Family: Leptospiraceae
Genus: *Leptospira (L. interrogans)
Family: Spirochaetaceae
Genus: *Spirochaeta (S. plicatilis)
          Brachyspira (B. aalborgi)
Family: Treponemataceae
Genus: *Treponema (T. pallidum)

Leptonema (L. illini)

Borrelia (B. anserina)
Crispispira (C. pectinis)

The rickettsias

Order: Rickettsiales
Family: Anaplasmataceae
Genus: *Anaplasma (A. marginale)
          Eperythrozoon (E. coccoides)
Family: Bartonellaceae
Genus: *Bartonella (B. bacilliformis)

Aegyptianella (A. pullorum)
Haemobartonella (H. muris)

Grahamella (G. talpae)

Family:	Ehrlichiaceae	
Genus:	*Ehrlichia* (*E. canis*)	*Cowdria* (*C. ruminantium*)
	Neorickettsia (*N. helminthoeca*)	
Family:	Rickettsiaceae	
Tribe:	Rickettsieae	
Genus:	*Rickettsia* (*R. prowazekii*)	*Coxiella* (*C. burnetii*)
	Rochalimaea (*R. quinana*)	
Tribe:	Wolbachieae	
Genus:	*Wolbachia* (*W. pipientis*)	*Blattabacterium* (*B. cuenoti*)
	Rickettsiella (*R. popilliae*)	*Symbiotes* (*S. lectularius*)

Order: Chlamydiales
Family: Chlamydiaceae
Genus: *Chlamydia* (*C. trachomatis*)

The mycoplasmas

Division: Tenericutes

Class: Mollicutes

Order: Acholeplasmatales
Family: Acholeplasmataceae
Genus: *Acholeplasma* (*A. laidlawii*)

Order: Mycoplasmatales
Family: Mycoplasmataceae
Genus: *Mycoplasma* (*M. mycoides*)   *Ureaplasma* (*U. urealyticum*)
Family: Spiroplasmataceae
Genus: *Spiroplasma* (*S. citri*)
Genera of uncertain affiliation:
     Anaeroplasma (*A. abactoclasticum*)   *Thermoplasma* (*T. acidophilum*)

Protozoan endosymbionts

Genus: *Caedibacter* (*C. taeniospiralis*)
Pseudocaedibacter (*P. conjugatus*)

Holospora (*H. undulata*)
Tectibacter (*T. vulgaris*)

Miscellaneous genera

Ancalochloris (*A. perfilievii*)
Chloroflexus (*C. aurantiacus*)
Excellospora (*E. viridilutea*)

Bactoderma (*B. alba*)
Chloronema (*C. giganteum*)
Stibiobacter (*S. senarmontii*)

Index